D0536702

Doing Time in the Garden

Doing Time in the Garden

Life Lessons through Prison Horticulture

James Jiler

Illustrations by John Cannizzo

NEW VILLAGE PRESS / OAKLAND, CALIFORNIA

Doing Time in the Garden: Life Lessons through Prison Horticulture

All rights reserved. Copyright ©2006 by The Horticultural Society of New York. Except for brief portions quoted for purposes of review, no part of this book may be reprinted, reproduced or utilized in any medium now known or hereafter invented without permission in writing from the publisher.

Published by
New Village Press
P.O. Box 3049
Oakland, CA 94609
Orders: (510) 420-1361
press@newvillage.net
www.newvillagepress.net

New Village Press books are published under the auspices of
Architects/Designers/Planners for Social Responsibility.
www.adpsr.org

Design by Rita Lascaro
Black-and-white illustrations by John Cannizzo
Color photographs by James Jiler

Printed and bound in China
First printing, 2006

Library of Congress Cataloging-in-Publication Data
Jiler, James, 1959–
Doing time in the garden : life lessons through prison horticulture / James Jiler.
p. cm.
Includes index.
ISBN-13: 978-0-9766054-2-3 (alk. paper)
ISBN-10: 0-9766054-2-2 (alk. paper)
1. Criminals—Rehabilitation—New York (State)—New York—Case studies. 2. Criminals—Rehabilitation—New York (State)—Rikers Island. 3. Gardening—Therapeutic use—New York (State)—Rikers Island. 4. Horticulture—Study and teaching—New York (State)—Rikers Island. 5. Prisoners—Education—New York (State)—Rikers Island. 6. Recidivism—New York (State)—New York—Prevention. 7. Horticultural Society of New York. I. Title.
HV9306.N6J55 2006
365'.66--dc22
2006009093
10-digit ISBN: 0-9766054-2-2
13-digit ISBN: 978-0-9766054-2-3

Contents

Dedicated to
the hundreds of inmate gardeners who,
during their stay on Rikers Island,
left something positive behind
in the form of nature and beauty.

Foreword

SERENITY, and, yes, even hope are hardly the words one associates with a jail or prison. Yet there is a small two-acre patch of New York City's famous jail on Rikers Island for which those words are appropriate. This book is written to describe how one program can help change the term "correctional facility" from a euphemism to a reality.

Across the United States, hundreds of thousands of former inmates are being released each year from jail or prison. This book describes the GreenHouse project of the Horticultural Society of New York (HSNY), a program for those incarcerated, which prepares them to change their lives so that after release they will not return to the behavior that led to incarceration.

The program is no panacea. There is no certainty that it will work for all, nor certainty, particularly for those with serious substance abuse problems, that no relapses will occur. Having said that, it remains a highly sophisticated program that combines classroom and hands-on gardening with life lessons about teamwork, responsibility and nurturing (plants and each other) and, because of that, a program that remarkably reduces recidivism.

The HSNY GreenHouse project draws on many resources, mostly non-governmental. We are grateful for the support we receive from the New York City Department of Correction, the greenhouse and the land around it, the superbly trained, highly motivated, and carefully selected corrections officers who both transport the inmates to and from the project every day and remain present during the classes and work programs. Beyond that, many generous foundations and individuals support the program salaries and classroom materials, and many big-hearted nurseries on Long Island and in Westchester County contribute extraordinary plant materials.

What may be surprising are some of our other "off-island" partners. In 1997 HSNY began a collaborative partnership with the three public library systems in New York City. Our GreenBranches program now has professionally designed gardens at seventeen branch libraries in the City. Those gardens are maintained by former GreenHouse "students" who form our aftercare program—the GreenTeam. Another GreenTeam project is installing and maintaining gardens for New York City Housing Authority projects. We also partner with other nonprofits as well as private clients.

All of the above is not intended to be self-congratulatory, but rather to encourage other jurisdictions to find ways to make this "jail-to-street" program work for them. We do believe that an outside entity, such as HSNY, is an ideal catalyst to initiate and coordinate such a program in a neutral and unthreatening way. All jurisdictions, whether county or city, have public libraries, plant nurseries, garden clubs or horticultural organizations and, also a jail that can replicate the kind of magical program we have in New York. Surely, too, they can understand that serenity and hope need not be incompatible with jail.

Anthony R. Smith
President, Horticultural Society of New York

Foreword

WHENEVER I THINK BACK on my time as the Commissioner of the New York City Department of Correction, there comes always a flood of contradictory memories. I recall feeling—and continue to feel—a profound depression over the sheer numbers of poor and overwhelmingly minority men and women, many with serious health, mental health and drug addiction problems, who populate the City's jail system. Over 100,000 on any given year come to Rikers Island to await the disposition of their criminal case or to serve jail sentences of less than one year. One can't help but think—as this procession of the poorest New Yorkers comes in and out—that we can do something better for our most economically distressed communities to keep young people from landing in jail. Occasionally, inmates are so depressed or mentally ill that they attempt and sometimes succeed at committing suicide. The stories, not uncommon, of the abuse many suffer (especially the women) always distress me, as do the difficulties faced on the outside that include drug addiction and inadequate health care. It is not uncommon to see a mother and daughter both locked up in the women's jail. Despair is easy in the face of these seemingly intractable problems.

On the other hand, Rikers Island is also populated with corrections officers and staff who serve selflessly and heroically to improve the lives of those who find themselves in jail. It is populated with inmates who, despite overwhelming odds, turn their lives around through the force of either sheer will or, more usually, with the benefit of a program or mentor who can tap into these mostly young people's intelligence and excitement at rediscovering education or finding a craft they love. During my tenure as Commisioner I would occasionally teach in the culinary program on Rikers, and the interest and excitement of the students who were trying to master a skill was always incredibly moving. These, in the end, are the most powerful memories for me—inmates and staff engaged in some life affirming activity, going well beyond the usual "care, custody and control" mission of jails and prisons.

Nowhere is this more clearly seen than in the horticulture program for men and women at Rikers Island. Though I can't quite remember my precise reaction to the call from Tony Smith, President of the Horticultural Society of New York, in 1996 to suggest we begin such a program on Rikers, I imagine it was likely along the lines of "Are you insane?" Initially, I couldn't quite envision the connection between inner-city young adults, horticulture and landscaping. Additionally, the logistics of moving people in and out of jails and back and forth to the greenhouse and the gardens seemed challenging, to say the least. The more Tony and I talked, however, the more I could see this actually taking shape and working. Inmates would have a chance to be outside and working at something they would enjoy instead of staying indoors and doing either nothing or rote jobs like polishing floors. And, I myself had experienced, as a home-grown New York City resident who before getting a small backyard garden in my thirties had never planted a thing, that gardening, planting, harvesting and even composting were really enjoyable activities that transcended all backgrounds and demographic characteristics.

Tony was right. It was a great idea. The inmates who participated in the program had wonderful experiences. The staff that ran the program was incredibly dedicated. Great mentors and teachers developed programs outside of Rikers so that landscaping and horticultural skills could be further honed. Many former inmates supported themselves with the skills they learned from these programs and went on to work in the field. It was always wonderful to see, as I traveled around Rikers Island, the students in the program busily planting, seeding and landscaping despite being in the midst of the world's largest penal colony. Most of these program participants had probably never even thought about growing flowers, much less actually doing the sometimes delicate, sometimes arduous task of weeding, pruning and feeding. Yet, there they were—eager and anxious to get outside, to take care of their garden and participate in the miracle of watching beautiful things grow.

The GreenHouse project remains a relatively small program in a jail system that holds over 14,000 people every day. Yet for many, it changed their lives forever in profound ways. It not only demonstrates that these kinds of programs could be replicated all over the country, but offers further proof that many in our jails and prisons want to do good and make good. With relatively small effort and a dedicated staff of people like James Jiler who give tirelessly of themselves, many former inmates have discovered the pleasures of creating and nurturing beautiful things and have simultaneously learned practical skills. Perhaps more importantly, many have found that it wasn't just the flowers and plants that they were so carefully helping to stay healthy and grow. It was, of course, themselves.

Michael Jacobson
President, Vera Institute of Justice

Preface

RIKERS ISLAND is the largest jail complex in the United States—a 415-acre island in Flushing Bay, ringed by razor wire and encompassing ten different jails. On any given day it is home to up to 20,000 inmates, a population that constitutes the bulk of New York City's criminal justice system. Despite its foreboding nature, a majority of inmates, 65% according to statistics, return to Rikers each year, caught in a cycle of recidivism and crime.

Since 1996, The Horticultural Society of New York has developed and administered a jail-to-street horticulture program for men and women inmates at Rikers Island, specifically to break this cycle. At our two-acre facility, complete with a greenhouse and attached classroom, inmates learn about plant science, ecology, horticulture skills, gardening construction and design. Apart from our instructional gardens, we build bird and bat houses and bird feeders and grow plants for gardens in the schools and parks of different neighborhoods, using our Rikers GreenHouse as a resource that helps restore the connection between people and nature in the City's urban communities.

Once released, students have the opportunity to work with us as paid interns, honing their skills as horticulturists as they build, plant and maintain gardens in New York City. The provision of jobs is exceedingly important for ex-offenders, for it keeps them productive, employed, and learning as they negotiate the difficult task of re-integrating back into society.

Often I am approached by staff at jails and correctional facilities in states and counties across the country and Canada and asked how one establishes an effective program for at-risk or incarcerated men and women. How do you engage people in developing job skills when they have no interest in learning? What kinds of hands-on projects are most successful? How do you determine what success is in the jail or prison setting? How do inmates find jobs after their release?

The answers to these questions are many and reflect the complex nature of different city and state facilities found throughout the country as well as the social dynamics of the people incarcerated there. At Rikers Island, for example, there are ten jails housing ten different populations of offenders. Unlike state programs, which can have inmates for extended periods of over a year, Rikers is for short-timers; many of our students are with us for less than six months, creating a constant turnover of new and old students. There is no one formula for a successful program. There are, however, a number of approaches one can take in establishing an effective program and applying it to the physical and social conditions at hand.

Doing Time in the Garden was written to inspire and assist practitioners in developing their own programs by sharing the collective experience and insights from our work on Rikers Island. While intending to be instructive, we also see this as the beginning of a dialogue between gardeners, counselors, horticulture therapists, correction officers and students engaged in horticultural work, both inside and outside of jail.

CHAPTER ONE

The Longest Bridge

Twenty-five hundred men saddled with an aggregate of twenty thousand years. Within such cycles worlds are born, die and are reborn. That span has witnessed the evolution of the intelligence of mortal man. Twenty thousand years in my keeping. What will they evolve? Will they bring life and purpose to any of our 2,500 men who are sharing that tremendous burden? It is hard to say. —WARDEN LEWIS LAWES

IT IS COLD as the car ascends the Francis R. Buono Bridge and crosses a narrow stretch of Flushing Bay to New York City's sprawling jail complex on Rikers Island. The bridge is more than just a well-constructed link between the Borough of Queens and Rikers Island. Only two miles long, inmates call this "the longest bridge in the world," one, they claim, "takes minutes to cross over but years to cross back." From its 100-foot rise above the water, there are sweeping views of Flushing Bay, the runways of LaGuardia Airport and the gleaming skyline of Manhattan. As the bridge dips and Rikers approaches, one has the physical sensation of leaving a familiar place for a secretive and altogether separate world—one surrounded by water, double rows of electronic fences, metal gates and razor wire.

A cold wind whips off the bay and hammers the car as it pulls to the gate for final clearance onto the Island. Less than five miles separate Rikers from mid-town Manhattan, but despite its proximity, the weather lags weeks behind the more spring-like conditions found in the city. The island, one of the largest penal colonies in the world, rests exposed in the bay like a floating barge subject to the wind and biting cold that sweeps off the water. Drifts of snow are piled on the curbsides and along the median of Hazen Street, the Island's main road. Yet in scattered patches on the median and by the entrance of the island's jails, tips of green stalks are emerging—the first hint of daffodils and crocuses, which, each year, like silent troubadours, herald the arrival of spring and the hope of new life.

The welcome to Rikers Island. Only designated vehicles, inmates, and persons cleared for security are allowed to venture past this gate and onto the bridge.

Few people entering Rikers bother with the subtle signs of nature, least of all the inmates, most of whom—apart from their allotted hour in a trampled grass yard—rarely see the outside of jail. But each working day for the past eight years I have crossed the bridge onto Rikers Island to do exactly that: engage inmates with their natural world by growing plants and building gardens. Enclosed within a metal maze of gates, fences and razor wire is a greenhouse surrounded by two acres of cultivable land where I teach horticulture to men and women inmates. The idea is to provide prisoners with job and life skills, some scientific knowledge and on-going therapy working with plants and animals in the hope they can redirect their lives through meaningful work. Our program, which includes hiring inmates or helping them find jobs after their release, aims to break the vicious cycle of recidivism, by most counts as high as 65% in the New York City jail system.

I give my clearance number to the officer manning the last gate onto the island and am waved through, passing the median strip on Hazen Street on one side, and ARDC—the Adolescent Reception and Detention Center—on my right, which, as its name implies, houses up to 2,500 adolescent males. With me is John Cannizzo. An artist, gardener and former book illustrator, Cannizzo is employed by HSNY as a gardening instructor, carpenter and job coordinator working with inmates both on and off Rikers Island. In his previous life, John worked in Kenya building water systems, taught art at an inner-city high school in Brooklyn, was a job counselor for refugees, and ran his own publishing company; for the inmates, he's their lifeline as they leave the greenhouse and return to their world outside of jail. His role is critical, for while many inmates harbor solid dreams and aspirations inside jail, once released, they are faced with the same issues that led to their arrest. Joblessness, a lack of education, a history of substance abuse and time in city shelters are just a few of the obstacles inmates deal with when they leave jail seeking to establish a new life.

John is a one-person after-care program. With persistence and a telephone he connects inmates to agencies that provide housing or drug care, or places them directly in jobs, in school, or in vocational training programs. He helps them secure identification or a driver's license, write a resumé, clear up old student loans or secure new ones. He gets to know their mothers, sisters, husbands, wives, children and employers. Every three months he calls to see how they are doing, whether an agency followed through with housing, or whether a job was offered after he'd set up an interview. When a student disappears, or fails to come to work, John hears about it, usually from a spouse or mother. After-care is more than a service: it is a relationship that is crucial in helping released inmates find a new path in life that keeps them out of jail.

We pull onto Shore Road, a narrow two-lane drive that hugs the steep coastline on the Island's northern edge, and park by a row of trailers perched on a small cliff above the water. The trailers house various administrative offices for the Department of Correction, including the Ministry and its contingent of priests, rabbis, sisters and imams who provide religious services to the city's inmates; the Education Department, responsible for overseeing ten jail high schools and educational programs; and the Department of Investigation, which serves as the City's watchdog for criminal activity inside the jails.

The view looking outward is impressive. Rikers Island sits at a point in Flushing Bay where Long Island Sound begins to narrow as it dramatically transforms itself into the East River, a misnamed tidal wash that links Manhattan harbor with the southern shore of New England. Pleasure crafts, barges and tugboats skirt the island well beyond the five hundred yards that are off limits to non-department craft. From Vinnie C., an officer and fisherman who runs a fishing boat on his days off from jail, I'm told that the best lobstering on the Sound is right off Rikers Island. "Because it's off limits to fishing," he said, "the lobsters here are big and there's lots of them." Every few hours or so, a Department of Correction craft patrols the water around the island, both as a deterrent for any inmate foolhardy enough to try and swim against the bay's stiff currents, and for boats veering too close to the Rikers shoreline.

In the distance, the Whitestone Bridge sends a graceful arch across the Sound connecting the borough of Queens to the Bronx. Co-op City, a high-rise development built in the 1960s for middle-income housing, looms to the north, rising from the Bronx marshlands like mushrooms in a plain. Just across the Bay is the jail barge, a huge, metallic blue floating prison that was bought to house the overflow of inmates who swamped the City's correctional system in the early 1990s. Built in New Orleans at a cost of $161 million, the Vernon C. Bain Center—named after a popular warden who was killed in a car accident—has 100 cells and 16 dormitories and can hold maximum and medium security inmates. With crime down across the city, the barge has become almost empty, resting now on its moorings like a floating ghost ship.

Above the island, planes circle noiselessly before their final descent into La Guardia Airport, literally a stone's throw from the southeast shore of Rikers. As they land, few passengers realize that the squat, star-shaped buildings rushing beneath them constitute a vast penal system. While Rikers has a vivid place in the public's imagination through television shows such as "Oz," "NYPD Blue" and "Law and Order," it is in fact, a hidden city-within-a-city with full services for up to 20,000 inmates seven days a week. This includes 60,000 meals a day, a high school in every jail, doctors, psychologists and nurses, a nursery for babies under one year of age, a fleet of buses to ferry inmates on and off Rikers Island, a harbor patrol, a canine unit, and over 4,000 officers on staff 24-hours a day.

Ask many native New Yorkers where Rikers is and they usually point to Randalls Island, a block of land in the East River facing northern Manhattan—home to a city psychiatric hospital, playing fields, training centers for the Fire and Police Departments and field headquarters for the Parks Department. Others mistake it for Hart Island in the Bronx, better known as

Potters Field, a place where the city buries its unclaimed dead using inmate crews from Rikers.

Synonymous as it is with the term "jail," many residents don't realize that Rikers Island is an actual island, owned by the City of New York since 1884, when it was purchased for $180,000 from the Ryker family and used as a farm to feed the city's jail population. The first jail was built on the island in 1936, and filled with inmates transferred from the city penitentiary on Blackwell's Island (known today as Roosevelt Island) in the East River.

A complex of 10 jails covers Rikers Island, which sits in the East River between Queens and the Bronx, adjacent to LaGuardia Airport runways. Aerial *(photo: Bluesky International Ltd.)*

Over the next forty years, dredging and landfill increased the size of the island from 87 to 415 acres. Additional jails were built, creating a large complex of separate facilities including a sentenced men's penitentiary, a woman's jail, a jail for adolescents, two maximum security facilities, a contagious disease section and a small prison hospital. In addition to housing inmates, Rikers, over the years, has run a nursery for the Parks Department, grown trees for the U.S. Forest Service, tended land as an in-house farm, and in the late 1960s, briefly served as a practice field for the New York Jets football team.

The greenhouse and gardens sit on almost two acres of land inside a double row of fencing adjacent to the Rose M. Singer House for Women. There are two ways to approach the greenhouse. With proper clearance, one can simply enter the woman's jail, pass through a maze of corridors and gates, to reach a small side door which leads outside to a fenced-in pathway. The path leads to wide double fence gate and into the garden. Male inmates, for obvious reasons, are driven to the greenhouse on an access road that winds behind the recreation field of C-73, a medium security facility for men, and a row of plastic dome structures called sprungs—used to house inmates in a drug program—to the greenhouse corridor right behind the woman's jail.

Although the greenhouse can be seen from Shore Road as well as from the women's jail parking lot, only a handful of the 5,000 officers and civilians who work on Rikers ever visit it, and most hardly know it exists. First-time visitors entering the gates are visibly surprised at the extent of gardens in the complex. For inmates, it is an oasis.

Six years of creative work and input by over three hundred and fifty inmates have helped transform a flat weedy field into a labyrinth of different gardens, a small emerging woodland and a waterfall and pond complex. Nothing here is static. As the inmates come and go they add their own distinctive touches to the landscape. One inmate, with salvaged wood and rocks plucked from the shoreline, designed and built a raised gazebo, then added a waterfall that flows off its side into a constructed pond stocked with fish and turtles. A succession of women inmates designed and built a raised herb garden using Belgian blocks that were donated by a federal prison in Georgia. Brick and stone pathways connect gardens through archways and trellises constructed by another group of student inmates. A former rose garden—part of the old adolescent program—is now an oval-shaped butterfly garden consisting of small trees, perennials and flowering shrubs. Pathways and borders have undergone a series of changes as rocks were substituted for logs and driftwood, or stone and gravel for woodchips.

Each garden flows naturally into the other as if following a master plan, but many of the sites were simply designed and planted with what one designer friend calls the "point and shoot" method. For our purpose it works. With creative use of space, some design aesthetics, a knowledge of plants and with whatever materials are available on Rikers, the students work at filling the gaps between each garden, fine-tuning the sites already developed, and adding embellishing details to create a harmonious and continually emerging landscape.

The impact is more than visual. As habitats for birds, insects, and reptiles develop, as plants naturalize and spread through the gardens, as trees mature to form new structures and micro-climates, and as diversity increases, the garden transforms into a complex of interdependent plant, animal and human relationships. Our job is to help inmates understand and explore these relationships as a way of creating profound connection to both nature and their community.

Beginnings of Prison Horticulture

The penitentiary system as we know it today has only been in existence for about 150 years. Established in the early 19th Century by the Quakers in Pennsylvania, the "Penitentiary" marked an end to the torture, executions and beatings for public display that were typically meted out for criminal behavior. Though fraught with harsh, unsanitary and generally inhumane conditions, the system introduced the concept of "loss of liberty" as a basis for institutional punishment. Early prisons offered uniform clothes, poor food, solitary confinement and hard labor.

As part of their upkeep, inmates were required to work in assigned details that would make the jail as close to a self-supporting complex as possible. These jobs were in construction (early prisons were built primarily by the prisoners inhabiting them), maintenance and sanitation, tailoring, quarrying stone, (materials to build the prison), food preparation and farming. Growing food was a requirement for running a prison and early prison farmwork entailed long hours, severe conditions and

left: A Rikers greenhouse *circa 1959* where inmates tended houseplants and grew annuals for use around the island. (*Courtesy of the NY Department of Correction*) right: Inmates pruning trees in the USFS tree nursery. (*Courtesy of the NY Department of Correction*)

the steady demands of prison quotas. Despite this severity, prison officials recognized that inmates exposed to the outdoors, hard work and fresh food were less likely to cause problems in jail than inmates locked in their cells all day.

In the 1840s, social reformers joined with architects and engineers to transform the prison system—and the prison itself—from a filthy place of confinement into a clean and orderly institution that promoted rehabilitation of the individual. The idea was that a strict diet of work and military discipline would help turn criminals into law-abiding citizens. This optimistic belief in the new correction paradigm was readily embraced: as one prison chaplain insisted, "Could we all be put on prison fare, for the space of two or three generations, the world would ultimately be the better for it."

State and federal penitentiaries were often located in rural areas far from urban centers, on land that lent itself to crop production and animal husbandry. Like Rikers Island, which still contains the largest farm in New York City and produces up to 40,000 pounds of produce a year, many prison facilities were actual farms before they were bought to house corrections. Angola Penitentiary, a maximum security state facility in Louisiana, is a former plantation with 18,000 cultivated acres, the largest working farm in the state.

In addition to farming, landscaping and horticulture were adopted to soften the harshness of the prison environment. The notorious Sing Sing Penitentiary in Ossining, New York, became a model of progressive reform by the late 1920's with the construction of a hospital, library, classrooms and gymnasium. The crowning achievement, however, was the establishment of expansive gardens and landscaped buildings on the prison grounds, all done by inmates led by Charles Chapin, a former New York City newspaper editor serving a life sentence for the murder of his wife. Called the "Rose Man of Sing Sing," a name Chapin garnered for the 3,000 roses he planted in his gardens, he became an ardent bird enthusiast after he noticed flocks of birds roosting in the prison's trees, apparently attracted to his gardens. With permission from Lewis Lawes, the progressive Sing Sing warden, Chapin raised funds from former business associates and media friends as well as inmates to build a huge aviary on prison grounds, siting it among the plants he tenderly cared for. The aviary, completed in three years, sported a huge dome and had room inside where the inmates could sit and admire the struc-

The greenhouse at Sing Sing and one of The "Rose Man's" circular gardens that helped transform the prison yard. (*Courtesy of Guy Chelli*)

The Sing Sing aviary built by Chapin with funds raised from New York socialites and prison inmates. (*Courtesy of Guy Chelli*)

ture's resident birds. After Chapin's death in 1930, the aviary deteriorated and in 1946 prison authorities had it demolished.

Not far from Sing Sing in Westchester County, the Bedford Facility for women, opened in 1901, quickly established an ethos of prison reform that would influence the future of U.S. Corrections. Under the administration of Jail Superintendent Katherine Bemett Davis, Bedford required inmates to not only work, but attend school. Basic math, reading and writing were taught along with mechanical drawing, stenography, typing, chair caning, bookbinding, painting and carpentry. Davis gave singing lessons, the assistant superintendent taught gymnastics and the physician gave weekly talks on physiology and sex hygiene. In summer, a recreational instructor was employed for exercise and sports events.

Davis also stressed outdoor work and "fresh air treatment" as a way of both promoting health and producing food. Inmates milked cows, raised chickens, and slaughtered pigs to supply all their own milk, eggs, pork and a good quantity of their beef and vegetables. They also planted trees and maintained flower gardens in front of the facility's four cottages.

Today, in a not so subtle departure from Davis's reforms, the majority of state prisons have work details in a variety of tasks associated with landscape maintenance or food production. There are prisons with large commercial greenhouses where inmates grow bedding plants, vegetables, herbs and flowering perennials for either the facility or for public spaces in the nearby township. Pennsylvania, the first state to build a penitentiary and establish prison farms, now hosts a network of facilities involved with raising livestock and dairy herds, agriculture, and food processing and canning. Their products supply more than half the food for the state's prison system. Pennsylvania is not alone. Others including Virginia, New Hampshire, Massachusetts, and Florida have extensive systems geared toward reducing food costs for their prisons.

Angola Prison in Louisiana has a nine-hole golf course built entirely by hand by inmates. Used by prison employees and the local public, the course sports hazards, bunkers, and seeded fairways and is maintained entirely by an inmate grounds crew, most of whom are in for life, yet because of "good behavior" are allowed to work on the golf course.

Most of these positions constitute "work details" which inmates are assigned to as part of their obligation of "doing time." Details are typically labor intensive, offer little skill development other than the task at hand, and pay an obligatory wage of 25 to 50 cents per hour. They are in fact, cheap labor for work that benefits the institution. Rarely is there any effort to link work details with employment opportunities after the person's release.

Less often, are landscape or gardening programs offered that teach inmates professional skills which enable them, much like a prison mechanic or woodworking vocational school, to find jobs in that field when their sentence is finished. Some prisons use horticulture therapeutically to help inmates cope with issues in their lives such as anger, traumatic stress, substance abuse, and depression. The Rikers Island GreenHouse and post-release GreenTeam programs take vocational training a step further, incorporating an eclectic mix of garden therapy, science and English literacy, life skill development and job enrichment with programs for job placement once a student has served his/her sentence. The concept that working with nature can have a profound and lasting impact on behavior is manifest in the kinds of activities employed at the GreenHouse Program. Students are required to work, attend class, take tests, and maintain the facility and gardens, but they are also taught to observe their surroundings as managers of a dynamic landscape. They learn to make informed decisions and monitor the results of those decisions in order to understand the types of natural processes that occur in nature.

Crime and Punishment

In the year 2006, over two million U.S. citizens are behind bars, the majority incarcerated for non-violent crimes and, especially in the case of women, for self-inflicted crimes such as substance abuse and prostitution. Almost five million adult men and women are under federal, state, or local probation or parole jurisdiction. The numbers in simple terms are staggering: one out of roughly 150 Americans is serving time in jail or prison, while one in 31 is on parole, on probation or behind bars. Compare that number to the early 1970s when a total of 200,000 Americans were incarcerated approximately one in 1,000.

I am often asked why I would want to help inmates when they should be punished, rather than rewarded for committing crimes in society. My stock answer reflects a hard reality: Inmates, unless they are serving life-without-parole sentences,

The view of the pathway leading past the Rikers greenhouse in fall and winter. The pathways were built in winter 1998 and planted later that year.

are eventually released, and many return to the same economic and social conditions that led to their incarceration in the first place. Only now they also have to contend with a prison record. Statistics in the criminal justice system reveal that offenders who leave jail with job skills and/or education and are given a modicum of after-care services are more likely to break the cycle of recidivism than inmates simply released. If ex-offenders are provided options, they can begin to contribute to their communities instead of engaging in crime that debilitates families and damages neighborhoods. There are immediate economic advantages, too—for example, lowering recidivism relieves New York City taxpayers the almost $64,000 required to prosecute, defend and house each inmate annually. Despite these findings, at both the state and federal level, programs designed to rehabilitate inmates have been curtailed or eliminated including, in 1993, the opportunity for sentenced men and women to attend college through the federally funded Pell Grant. (In terms of reducing recidivism, the Pell Grant was one of the most well-documented success programs in the criminal justice system.) The result is a nation-

The view of the pathway leading past the Rikers greenhouse in spring and summer. The change of seasons reflects the dynamic nature of our gardens, which in itself becomes a metaphor for individual change.

wide $35 billion-a-year penal system that warehouses people rather than aids in their rehabilitation and return to society.

The GreenHouse Program on Rikers is a boon for the Department of Correction, since it costs the City almost nothing to run except for the in-kind presence of two officers who escort the inmates around the island and maintain security while they're at the greenhouse. Established and administered by the Horticultural Society of New York (HSNY), a century-old, non-profit organization based in Manhattan, the Rikers program seems somewhat unusual for an organization known primarily for the New York Flower Show and its blue-blood lineage. The Society, however, is no stranger to the jail, having run a similar project on the island for adolescent men from 1986 to 1993 with money from a city youth grant. When funding was cut, the project was terminated, and for four years the greenhouse lay vacant, visited by a few officers who kept the weeds in check with herbicide and a tractor mower.

In 1996, Anthony Smith, the newly-appointed president of HSNY, and a former Assistant Deputy Commissioner of the Department of General Services in the administrations of

Mayor David Dinkins, reinstated the GreenHouse Program with a well-placed call to Commissioner of Correction, Mike Jacobson, an old colleague in city government. A board member of South Forty Corporation, an organization that provides after-care services for ex-offenders, Smith is a big supporter of rehabilitative programs for inmates.

"I strongly believe that horticulture offers all the elements—education, job skills, self-esteem, creativity, confidence—of a successful program," said Smith. "Offenders need something positive from their jail experience. After all, it's called the Department of Correction, but too often it's strictly punishment. And much too often inmates keep returning to jail."

Smith recognized the difference between horticulture as a jobs skills program and farm and landscape labor as a grueling detail. Since 1981 Rikers has operated a vegetable farm encompassing six to eight acres scattered across the island. On 50 by 50 foot plots, up to one hundred inmates toiled in long rows, planting weeding and harvesting.[1]

"They could probably tell the difference between a tomato and an eggplant," Smith said to me. "But no one's coming out of there with the skills to get a job in Central Park or the Brooklyn Botanic Garden. I want our students to have the skills so when they leave Rikers they can find work in horticulture."

A mix of grants from private individuals and charitable foundations gave HSNY the financial support to hire staff and make the physical move to Rikers. But after years of non-use, the greenhouse was in disrepair. Electricians and a plumber had to be called in. The water pipes that carried steam heat had to be flushed and cleaned. The glass in several panes of the greenhouse roof had to be replaced. The Department of Correction needed to assign officers to the program from another tour of duty, a challenging process at Department of Correction, but with strategic phone calls and meetings we managed to secure one female officer for the women and a male officer for the men. More meetings ensued to find a bus to transport the men across the island to the greenhouse. This all took months, and my entry into prison life took place in incremental steps. I soon became accustomed to the daily rigors of driving onto the island—well-versed in the security risks of working on Rikers, used to seeing men in orange jumpsuits sweeping out the office trailers, cutting grass and moving across the island on flatbed trucks as they went about their work.

Inside the Rikers greenhouse a student tends tropical houseplants in winter. By spring the tables will be full of germinating seedlings.

Within a year the greenhouse was open for business. Fifteen women would attend morning class, followed by ten men in the afternoon. They would come as volunteers earning the minimum 20-cents hourly wage as part of a work program. Each group would work from four to six hours a day with instruction. Together, we would begin the process of transforming the bleak grounds of the Rikers site into a series of flowering gardens and vegetable beds, a woodlands and nursery. We would grow plants and build planters and

Seedlings germinated in the greenhouse are ready for the Rikers vegetable garden by the April 15th planting date.

birdhouses for under-served neighborhoods in New York City.

Effectively changing my students' lives through this process would remain for some time an ideal—a social theory based on the likelihood that inmates, despite all the problems associated with being in jail, and the many problems they were likely to face once released, would find transformation in nature. I had my doubts. Afterall, the majority of my students were from the inner-city, and the likelihood that planting a garden could make them stop the lucrative business of selling drugs or stem the hard-core addiction to taking them seemed remote.

[1] In 2001, the Rikers farm became more programmatic with work crews of only 6 to 10 inmates responsible for 3–5 acres. The farm works closely with the RIDE program (Rikers Island Discharge Enhancement) which focuses on inmate's needs as they prepare to leave jail.

CHAPTER TWO

Horticulture as Therapy

It is a fact that the three largest mental institutions in the world are the Los Angeles County Jail, Rikers Island in New York, and Cook County Jail in Chicago.

—ALAN ELSNER, GATES OF INJUSTICE

Plants for Mind and Body

SINCE THE EARLY EMERGENCE of humans on earth, the relationship between plants and people has been a profound and ever-evolving experience—one that has marked both our survival and development. It is no wonder that we gain solace working with plants and have used gardening to quiet the mind, reduce stress, and create a general wellness of being.

Horticultural therapy as a concept and practice saw its beginnings with the rise of institutional medicine shortly after the American Revolution. As early as 1798 the U.S. medical profession, led by Dr. Benjamin Rush, was publicizing accounts of patients who, after working in the hospital food gardens to help pay for their care, recovered more quickly than wealthier patients who simply convalesced in their rooms. Sanitaria and psychiatric hospitals began to adopt gardening as an in-house activity for their patients, not only to grow food for the institution, but also to promote healing of the mind and body. In 1897, Friends Hospital, a psychiatric institution in Philadelphia, built a greenhouse for the sole purpose of providing its patients with therapeutic horticultural activities.

Beyond the physical attributes of fresh air and exercise, gardening and plant care was seen to provide a greater range of benefits, from stress reduction to increasing self-esteem or improving mental focus to redeveloping motor skills affected by trauma.

Other rehabililative qualities are perhaps related to something less tangible but more profound—the deep mystery of nature, a force that in many circles is seen as a powerful vehicle for individual growth and development. Psychiatrist Karl Menninger calls this process *adjunctive therapy*, which can help patients with depression, anger and trauma disorders. The serenity involved with designing and constructing a garden, along with the work of maintaining it, can channel anger and aggression into productive skills and activities.

After decades of alternating interest and indifference in the medical community, horticultural therapy is again gaining popularity as a viable program for a patient's physical and emotional recovery.

How does this play out in the prison setting? Unlike programs in hospitals, drug abuse centers, psychiatric wards, or hospices that target select populations, prison populations encompass a range of personal and mental disorders, physical ailments and anti-social behavior. Recent studies claim that up to 75% of all Rikers' inmates have some form of mental illness. While national statistics are murky, studies carried out by the National Commission on Correctional Health Care conclude that inmates have incidence of mental disorder three to four times greater than that of the general U.S. population. In 1999, the Bureau of Justice Statistics released a report that estimated 283,000 inmates nation-wide were mentally ill, almost 16% of all people behind bars. For white women 24 years or younger, the rate of mental illness was almost 40%. While the mentally ill constitute a staggering percentage of those who return to jail, most U.S. correctional institutions do not address the underlying mental disorders that may have led to criminal behavior in the first place. Nor do governments work with the mentally ill once they are released. Of 547,800 parolees who were cited as being mentally ill, few were provided services unless the symptoms were a specific cause of their arrest. (source: Joan Petersilla, *When Prisoners Come Home.*)

With the prison population comprising such an eclectic mix of prisoners and patients, horticulture "behind bars" aims to explore the potential of each individual and his/her struggle for change, growth and self-realization. The garden provides an important medium for this dynamic process. What takes place in the garden—work, planning, skill, and an understanding of the various plant, animal and human interactions—has benefits that affect individuals in a myriad of ways. Plants respond to care, and a garden rewards the caretaker with food, beauty, flowers, and a positive response from the community (or from the professional staff or fellow inmates) that is not readily found in other settings or work sites.

For prisoners, many of whom have suffered frequent failures in the job place and the frustrations of being marginalized in society, horticulture is a process that allows them to control their environment through shared responsibilities in an unspoken contract between person and plant. Accomplishment is its own reward, generating new goals and productive efforts in a person's life. As skills develop and projects increase, individu-

The Gazebo and pond: Tibetan prayer flags fluttering from the roof deter the Island's wild Canada geese from using the pond.

als achieve a greater sense of empowerment, along with newfound pride in their role in the workplace.

Therapists with the Garden Project at the San Francisco County Jail use plant growth and the cycles of nature to emphasize natural and controlled processes of personal growth. Compost is used as a metaphor for life's mistakes and misfortunes and suggests that the individual can redirect their path in life from lessons learned. Weeding is equivalent to removing the negative thoughts, patterns of behavior and influences in one's life, while transplanting and watering symbolize the stage of leaving jail and maintaining a productive life outside.

The garden provides a direct hands-on method for individuals to take responsibility for themselves as a natural out-

Making a Difference through Gardening

By Catherine Sneed

Every year, in San Francisco and across the country, thousands of people pass through our country's penal justice system. If you work in this field, you know the numbers. Today's trends in criminal justice call for all the rest of us to give up on these people. The criminal justice system is based on the idea that there are people society has no hope of changing. Such hopelessness is not about our lack of faith in that person, as it is about our own lack of our humanity. How much we spend on our criminal justice system shows how much we have given up. Instead of a national trend, it should be a national shame, because these people never had a chance.

Students, proud of their work, strike a pose in the Rikers garden. Their time here will hopefully shape their outlook for a positive future.

I know the people I work with have lived in households that for generations have never seen a paycheck. I know most of them read and spell at the third grade level, though they may be in their twenties or thirties. Society fails them from the beginning—giving them a free ride to the city jails rather than the city college.

These are daunting facts. When I became a counselor at the San Francisco County Jail in 1980, I wondered how I could impact this cycle. My friend, the author and poet Wendell Berry, wrote: "Out of a history so much ruled by the motto Think Big, we have come to a place and a need that requires us to think little." The idea for this program began simply. When I first came to the San Francisco Sheriff's Department, I worked as a counselor. Everyday, I tried to convince the prisoners that just because they had done bad things, they were not bad people. They continued to return to jail. What I was telling them was not enough. It also was not enough to let them go with the clothes they came in and bus fare and expect them to change their lives. They needed to find a sense of purpose in their lives—a connection to something beyond themselves. I had just read *The Grapes of Wrath* by John Steinbeck and thought, if the families in that book could find hope in the land, so too could the prisoners. I had nothing else to offer them. I knew their lives were at stake and that I had to begin somewhere.

Beginning somewhere is enough: the economist E.F. Schumacher once wrote, "Perhaps we cannot raise the winds. But each of us can put up the sails so that when the wind comes we can catch it." Since 1982, I have been director of a gardening program for prisoners in the San Francisco County Jail and a post release program called The Garden Project. It is not a huge program, although over the years we have worked with thousands of individuals. During their time with us, they learn many things. They learn to grow food and to plant trees. They learn to be at work on time and to respect others. They give back to the communities they had taken from, and they earn a paycheck. There are other things they learn—things I cannot teach them—that transcend their day-to-day work and give them back their lives, and their humanity. Since our post-release program began in 1992, we have literally changed the face of San Francisco's poorest neighborhoods: we have planted over 7,000 trees, and the vegetables we grow feed hundreds each week. The Garden Project works: Garden project participants are less than 25% as likely to return to jail as non-Garden Project participants. So we don't just grow plants—we grow people too. We've put up our sail. ■

growth of taking responsibility for a garden. The more inmates play a part in the garden's creation and maintenance, the better they are able to apply these concepts to their lives. In the process, inmates develop important job skills, including construction, gardening and landscape maintenance, interior design, floral arrangement and general management abilities that can help them find employment when they re-enter their communities. Often, gardening programs are based on curricula that lead to certification in different aspects of horticulture, including IPM (integrated pest management), pest control, turf management, tree care, landscape maintenance and plant propagation.

I have found that working one-on-one with inmates in the garden is the most effective way to address some of the issues affecting their lives. Unlike the formal groups that inmates are often assigned to that address substance abuse, anger management, or life skills, our informal approach is effective in getting inmates to talk about themselves. The act of gardening is a form of therapy, and through it we constantly explore each other's values, goals and life objectives. While I do not presume to have the skills of a professional therapist, working side by side, I've encouraged students to seek a therapist's help, to enroll in school or a drug program, and in more general terms, to exploit their strengths in planning for the future. Often it's important for the students to just talk about their past in an honest, open manner without the judgment of a formal group encounter. One former student expressed the sentiment of many when she said, "Being away from *the group* is by itself liberating. The one place in jail we can find that is here in the garden."

Yet to work in the garden is a privilege in itself. Inmates are given specific tasks and responsibilities equivalent to holding a job. They cannot have more than three unexcused absences.

Horticultural Therapy Works in Reducing Crime: A Program in Paris, Texas

In Paris, Texas, 6th District Judge Jim Lovett set a new course for probationers under the Lamar County Corrections Department. In an effort to reduce recidivism while reimbursing the county for the crimes its taxpayers had to bear, Lovett approved a new program based on horticultural therapy. To determine the effectiveness of the program, the Department, headed by Gary Marlowe sought out the help of Dr. Richard Mattson of Kansas State University, a leading authority on horticulture therapy, who agreed to establish a basis of research. Beginning in February 2001, and continuing for three years, 383 probationers were tested through four phases of the Horticulture Therapy program while carrying out tasks such as environmental restoration, food production, and public space beautification. They were compared with probationers involved in more typical community service activities such as trash clean-up, cemetery maintenance and janitorial work.

The results clearly showed the benefits of horticultural therapy; those in the horticulture test group showed significant changes in self-esteem, general and specific horticultural knowledge and a high environmental awareness. More importantly, those who took part in the twelve-month horticultural therapy program had a recidivism rate of 26% compared to 49% for those parolees not involved in therapeutic activities. ■

They are expected to be attentive in class, take tests, and carry out their designated assignments. They must get along with their colleagues, which often requires a high level of problem solving. For many, it is the first job they ever held, and more important, worked to hold.

Rikers Island: Doing Time in the Garden

BY 8:00 AM, Officer Ross has already brought the women to the greenhouse. A tall, large-boned woman with sixteen years in the Department, Ross spent much of that time with the Rikers Farm Unit before her assignment to horticulture. Her job, apart from the official duties involved with maintaining security, is a hybrid mix of law enforcer, social worker, counselor, judge, chef, horticulturist and den-mother. Ross escorts the women from inside the jail to the greenhouse. She screens applicants for their interest in horticulture and their eligibility to work outside. (All inmates are assigned classifications based on the severity of their crime, and their behavior inside jail. An inmate with a high classification would not be permitted to leave the inside of the jail.) She keeps order in the greenhouse, advises, chastises and gives inmates different tasks apart from their daily assignments in the garden. The greenhouse facility is Ross's personal domain and she presides over it like a kindly despot. Break the rules, smuggle contraband materials (anything not jail-issued is contraband) from the greenhouse into jail, fight, refuse to work, act insolent or disrespectful, and you are banished from the garden. Less severe offenses, like sitting in her leather armchair, is accompanied by a sharp reprimand. So is swearing. And kissing.

Ross is also protective of the students. "Work hard and good things will come of it," she tells them. If a former student is re-incarcerated, Ross finds a way to bring her back to the greenhouse. She secures decorations for Halloween and Christmas, recipes for cooking eggplant, extra jail-issued clothing or supplies such as milk, sugar, cereal and coffee. She has firm advice for students in abusive relationships—"You gotta leave him. Yesterday!" or for women battling Social Services for custody of their children—"Get straight. Stay straight. Find a job. You'll get your kids back."

The perennial garden designed and built by men students in 1999 for the classroom building.

"Life with all its complications," Ross tells me, "can really be quite simple."

Thirteen women are sitting in the classroom adjacent to the greenhouse when John Cannizzo and I arrive. The classroom is a brick and concrete structure that for all purposes

A student enjoying a spring moment in front of the gazebo.

serves as an office, library, kitchen, and workspace for our time at the greenhouse. There is a refrigerator and stove, three desks, two computers, shelves with a well-stocked library of gardening books, supply cabinets, a blackboard, a tropical fish tank and an aquarium where two turtles from the gazebo pond spend the winter. With the wind blowing on this cold morning, the classroom is a warm respite from the walk outside.

The faces of the women assembled look no different than any group of students in a learning class for adults. In their green and orange jump suits, they are inmates—women serving time for prostitution, drug possession, drug sales, assault and occasionally homicide. Yet, from the moment we begin work in the greenhouse or garden, they shed their roles as inmates and become my students. They are mothers, daughters, workers, wives and homemakers. They are women with a past that has little to do with the crime they're serving time for. Some have apartments, families and jobs waiting for them when they leave jail. Others can expect only a bed in a shelter and maybe a program to help with their addiction to drugs. A number of my students have life-threatening ailments. Others are embroiled in long and agonizing fights to reclaim children who were placed in foster care. During our time in the garden, we will trade stories, laugh and even argue. We will talk about our interests, the mistakes we made, and what we hope for the future. Inside the greenhouse anything is possible: it is fertile ground for dreams, a place where new jobs, educational opportunities, renewed family ties, and healthy relationships sprout like mushrooms from the decaying matter of old lives. The challenge lies in taking the dreams achieved inside back to the hard reality of their lives outside. In all our conversations rarely have I heard a women student voice self-pity.

While most of the students spend the first minutes in the greenhouse drinking coffee and decompressing from the stress of life inside jail, Denise R. heads outside the classroom towards the shed where we store the tools. Denise is in her late twenties, strongly built and, despite living most of her life in the hot Caribbean tropics, is hardly fazed by the March cold. Rather, she is grateful to be at the greenhouse relieving her stress with hard labor. A mother of six, including a two and three year old, she tries to stay busy, for downtime is "slow time" accompanied by a flood of thoughts and worries about her kids at home. "Gotta stay busy," she says. "Otherwise you think too much." Grabbing a shovel and spading fork, she heads towards the back of the greenhouse facing Shore Road and two rows of razor wire fencing. There she plunges into the task of carving beds into the raised herb garden. She makes a circle with Belgian blocks to form a center. Three-foot beds shoot outwards like rays from a sun, and in between, separating each bed, is a brick pathway.

Denise's work is part of a process that began the year before when a student first designed on paper her idea for a garden. When that student left another took her place, slightly altered the design, drew it to scale and with her group, broke ground and created beds out of salvaged bricks, two-by-fours and plastic edging. Now Denise has re-designed the beds and, using the donated Belgian blocks, is building a more permanent and elegant border. Soon, I tell her, we'll plant lavender,

The newly built and planted herb garden with Belgian blocks donated by a Georgia penitentiary shortly after completion in 2000. String around the perimeter keeps out hungry Canada geese.

sage, thyme, basil, mint, roses and rosemary. By summer we will have herbs to make soap, hair conditioner and hand lotion, products that are contraband in jail, but are used judiciously, and with supervision, at the greenhouse.

Jackie A., a sturdy Hispanic woman, serving a year for assault, heads immediately from the classroom into the greenhouse. A florist in her husband's shop before her arrest—she stabbed a woman in self-defense—Jackie is the de-facto "manager" of greenhouse operations. With a calm efficiency, she produces a batch of seed-starting medium, measuring out three parts peat moss to one part perlite and vermiculite in a large metal bowl, spraying it with water, then mixing it together by hand.

The thirty-by-sixty-foot greenhouse with its three long worktables is not exactly a model of technology or efficiency. Several panes are missing glass and the heater occasionally crashes, leading on cold nights to the demise of tender seedlings and tropical houseplants. Too much heat can also kill the plants, especially on weekends when the greenhouse goes unattended for several days. Because our access to the greenhouse is limited to the hours our officers work, we can't monitor troublesome conditions; yet, despite the sudden frosts and heat waves, we do remarkably well. The greenhouse produces an annual yield of several thousand vegetable, bedding and perennial seedlings that are used on Rikers or distributed to elementary schools, libraries and community gardens in the City.

Working with Jackie is Sonia B., a young Dominican woman, who quipped, when I first showed her the greenhouse, "they should lock me up for all the plants I've killed." Under Jackie's direction, Sonia has learned the fundamentals of running a greenhouse, and in the process has discovered both a talent and penchant for growing plants. She and Jackie are from the same neighborhood in the Bronx and as acquaintances from the streets, make a formidable team in the greenhouse. Sonia has eight months on her sentence—the result, she says, "of selling the wrong thing to the wrong guy." Twenty-four years old, she talks of changing her life, first by kicking her habit with drugs and completing her GED, then finding a job that pays enough for her family to live on. Finally, Sonia hopes to regain custody of her year-old son who, through a court order, is living with her mother. With her charge of possession, Sonia is just one of 56,000 women inmates in the U.S. jail system incarcerated for substance abuse, and her son yet another child who will not spend time with his mother during her rehabilitation.

Sonia's is a story that is replayed throughout prisons and jails across the nation. The population of female inmates has tripled in the past decade, and 75% of incarcerated women are mothers. In 1991, the National Council on Crime and Delinquency estimated that on any given day there were 167,000 children in the U.S. whose mothers were in jail, three-fourths of whom were under the age of eighteen. Studies also show that children of jailed parents are often traumatized by the experience, and five to six times more likely to end up behind bars.

Jackie hands Sonia a tray filled with seed mix and tells her to plant *echinacea*, a perennial commonly known as cone-flower. When ground into powder its roots are believed to boost the immune system. Jackie begins misting rows of germinated seedlings stretched out in assembly-line fashion down the long greenhouse benches. "The lettuce," she says, "is ready to go."

Inside the classroom, Laura M. is sitting at a table designing the vegetable garden the inmates will build and later plant. She carefully measures out the beds on graph paper, each grid representing one foot that will transform the nursery into a productive site for food. All winter the nursery was used to contain shrubs and small trees that were donated by different nurseries in the New York area and heeled in for the cold winter months. In early spring, the plants are dug up and donated to a number of gardening projects in the community. HSNY's GreenBranches program, a project that designs and builds gardens for branch libraries in low-income neighborhoods, is a chief recipient of greenhouse plants. The nursery space is temporary. As plants leave the nursery, the soil is turned over and

Rikers compost, a concoction of leftovers from 60,000 daily inmate meals and woodchips processed in a state-of-the-art sanitation facility constructed on the island, is spread a half-foot thick over the entire site. By April 15th, the official end-of-frost date, the garden will be ready for a spring crop of lettuce and broccoli.

Laura, twenty-two years old and serving an eight-month sentence for smuggling narcotics, is best described as a cross between Martha Stewart and Mariah Carey. She is not only an excellent and devoted gardener, but also draws, paints, and arranges flowers. She is a talented singer, songwriter, and actress, the combination of which earned her a living as an exotic dancer for strip clubs in Manhattan. "After a week at my job," she tells me, "the girls had me choreographing all their dances." Laura is passionate about animals, rattling off the names of her dogs, cats, snakes, fish, birds and lizards like members of some surrogate family. She recognizes that stripping is a limited career choice. In jail, she is making plans to enroll in a two-year program at a community college in Queens to become a veterinary technician.

"It's only a beginning," she says. "Eventually, I want to be an actual vet."

She peruses a book on landscape design and carefully draws the outlines to the garden. The artist in her avoids monotonous rows of planting beds for something more aesthetic. Inside a seventy-by-thirty-foot rectangle she has drawn a circle. From the edge of the circle, beds spread out in concentric patterns like a mandala in three dimensions. A network of interweaving pathways creates space between the growing areas and the garden fence line.

"Joanne is back," she tells me, coloring a pathway. Joanne B., one of my more dedicated students, was released a month ago. As a "first timer" in jail, with a job as a medical technician and a home to go back to, Joanne was statistically one of the few who was least likely to return to Rikers. "Joanne saw her in commissary."

Apparently, according to the "jail mill," Joanne angered her boyfriend when he saw her with another man. He called the police claiming she stole a gold chain from his apartment. They promptly arrested her for violating parole.

"But if she's innocent and they didn't find the chain," I said, "how can they arrest her?"

"Doesn't matter," one of my students tells me. "Any contact with police is a violation. By the time they hear your case you're already doing time."

"It's cynical," another says. "They let you do short-time knowing you're going to violate parole and head back to jail. They can nab you for just talking to a felon. 'I didn't know he's a felon,' you say. They don't care. They got you," she says. "Don't cop to short-time," she tells me. "You'll end up doing all of it." I thank her for the advice. I inquire whether we can bring Joanne back to the greenhouse.

A student cleans the pathway near the herb garden. Each student is responsible for maintaining a designated garden area.

"She'll be gone," Laura says. "They're packing her upstate. I think she got a year." She turns the paper and starts labeling the plots containing the plants she wants for the garden. "Look," she says. "I'll put the eggplants here mixed with squash, and the bush beans here with marigolds and tomatoes." On the corner of the page she sketches a small rendering of the garden. A bean teepee rises from the center of the mandala. "I'll be gone by summer," she says, as if an afterthought. "The food from this garden is for everyone else."

There lies the challenge for programs like ours. Unlike state or federal penitentiaries where inmates serving long-term sentences can become long-term gardeners, Rikers is a city jail serving short-term sentenced offenders. Most of our inmates spend an average of six to seven months at the greenhouse. As they leave, Officer Ross screens new students to take their place. Our program for men proceeds in a similar fashion. This revolving door results in different students at different skill levels entering the program at all times. Because students typically work only two to three seasons, we rarely expect them to leave as accomplished horticulturists ready to assume professional jobs in the outside world. Often a student only sees a partial completion of their project or a seasonal glimpse of the garden. While this has its drawbacks, the quick turnover allows us to expose a greater number of inmates to the world of

Summer maintenance in the bird and butterfly garden. Initially a rose garden in the 1980's and the only developed space when the program resumed in 1997, it is now a rich mix of perennials, deciduous shrubs and small flowering trees.

plants and gardening. A few will be stimulated enough to pursue a career in horticulture. Others use their time in the garden to examine their life and plan for a more promising future after their release. And a few will return to jail. But the common thread running through our students' lives and tying their experience together is a strong connection to the garden. It stays with them long after their release from Rikers. Through trauma and success this connection to nature resides as a heightened awareness of the world around them, and therein, themselves.

Spring clean-up in the Rikers Gazebo garden adds moments of personal reflection.

In jail, time is a double edge sword: inmates live for its passing, yet they are constantly aware of its presence; a six-month stint simply becomes 180 days and a "wake-up." This focus saturates its passage that, in the words of one student, "slows hours into days and days into months." But time tends to slip away in the garden. The process of transforming a barren piece of ground into a flowering garden, or taking something overgrown and weedy and restoring order and beauty is not measured by time. Work becomes a meditative process rather than a series of dreary tasks. Gardening becomes an avenue of self-expression, and through the accumulation of knowledge—empowerment.

Grabbing several pairs of hand pruners, a hand saw and a lopper, I take three new students to the cold outside. The day before I had given them a lecture on pruning shrubs. "Rule of thumb," I said. "If the shrub flowers in the summer, then prune it in late winter, early spring. If it flowers in the spring leave it alone until the summer. This way you're not taking off flower buds that were set last fall."

Three large Buddleias, well over six feet tall grace the entrance to the bird and butterfly garden. “These set their buds in late spring, early summer,” I tell them. They can be lopped short, about two to three feet from the ground.” I show them where we cut back the year before and explain to them the notion of shape and balance. “Give it a try,” I say. The women, in their large orange jumpsuits, orange jackets, and orange hunting caps, circle the shrubs warily, like boxers sizing up a formidable yet immobile opponent. One of them jabs with her loppers. A branch falls. The others begin jabbing and cutting. “Shape and balance,” I tell them.

And so it begins on a cold spring day—fourteen women immersed outdoors in the simple act of planting, pruning, turning the earth, spreading compost, marking borders, sowing seeds, even if the outdoors is surrounded by double strands of razor wire, metal gates and cyclone fencing—as our students do their time in the garden.

CHAPTER THREE

Jail and Community

One of the biggest failures of our justice system is that we put people in jail, but from jail we don't turn people out any better. We generally turn out people who are worse, and we're sending them back into your community.

—SAN FRANCISCO SHERIFF MICHAEL HENNESSEY

A Resource for Communities

The concept of rehabilitation is directly related to the role correctional facilities play in society, and how it is manifested has a profound impact on communities across the country. Since the Code of Hammurabi, 1795 BC, which institutionalized the idea of retribution for criminal acts, incarceration has swung back and forth between being punishment and rehabilitation. Today, when "tough on crime," carries strong political weight, there is less emphasis on prison alternatives or rehabilitative programs and more on the concrete construction of prisons. These large, hugely expensive facilities are what an ecologist might call a resource sink, a system that takes in substantial resources but gives little back. Indeed, prisons absorb vast public funds for facilities and services yet their value to society is to simply warehouse people who have committed crimes.

Along with other states such as California, New York expends more public money for prisons than it does for higher education. Critics of the state's criminal justice policy system argue that if a fraction of those costs were channeled into inner-city schools for smaller classrooms and after school programs than there would be less need for new state prisons. After all, the majority of men and women incarcerated in New York are from inner-city neighborhoods in the state's urban centers. At Rikers it is even more telling: 75% of all inmates come from five main neighborhoods in New York City. These demographics are not just limited to New York. It is well documented throughout the country that people released from jail are likely to move into core areas of impoverishment where housing is cheap and available.

Jails and prisons are designed to physically isolate the individual from society, and in New York, prisoners are shipped to the far reaches of the state, typically in rural counties, making it more burdensome for their family to visit on a regular basis. The difficulty of re-integrating back into society is magnified with years of segregation from family members and the social norms of a community or neighborhood. Researchers studying the problem of re-entry often cite this prolonged detachment as one of the primary reasons why job training and job placement programs may not work as well as expected in reducing recidivism.

At Rikers, our program is focused on creating as many opportunities as possible for inmates to connect with their community from jail. Surplus food grown in our vegetable and herb gardens is distributed to homeless shelters and soup kitchens in Manhattan. Plants are grown and propagated in the greenhouse for elementary schools, community gardens, day care centers, libraries and other public institutions. Inmates are taught rudimentary skills in carpentry as they build window and planter boxes, benches, and trellises for public schools and libraries.

Often these projects receive targeted funding and therefore have specific requirements. For example, butterfly habitat gardens made of planter boxes were constructed at Rikers, then re-assembled and planted on the roof of P.S. 96, an elementary school in East Harlem, by former inmates together with schoolchildren. Materials and labor were funded by a grant from the

Planter boxes constructed at the greenhouse made their way to a Manhattan rooftop owned by the Bridge—an organization that provides services for the mentally ill. Additional construction by GreenTeam interns and Bridge clients followed by a planting transformed the site into a private garden.

New York Community Trust and the City Gardens Club of New York. The success and growth of our projects are dependent on partnerships with different organizations throughout the city. There is no limit to the kinds of projects we carry out or who we partner with, as long as we can help inmates connect to nature, the community and their own creative powers.

In winter 2002, using old fire hose donated by the city Fire Department, we designed and constructed heavy-duty hammocks for the leopards and mandrills at the Staten Island Zoo. The project, described as "a way to increase vertical living space for zoo animals in small enclosures," was a metaphor readily grasped by the inmates building the hammocks. At the same time we began Project Jailbird, an educational program that, in partnership with a non-profit called City Beasts, brings non-releasable injured wildlife into jail so that inmates can experience urban wildlife up close. Raptors such as kestrels, owls and hawks, as well as reptiles and prey mammals make their way to Rikers several times a year, which educates our students and acquaints them with nature.

In return, we construct nesting boxes for owls, bats and kestrels at the greenhouse. These are then distributed to school and park gardens in the city to improve habitat for important urban wildlife. We also monitor songbird populations that flock to the feeders in our gardens. The information is then fed into a National Audubon program called "Feeder Watch" which tracks songbird populations across the country.

A bat house designed and built at the greenhouse finds its way to a city park in Manhattan.

The connection between jail and the wider community is an important aspect of the rehabilitation process because it makes inmates feel like a productive part of society during their incarceration. Furthermore, it allows the jail to serve a larger social function than simply removing criminals from the general population. For jail officials, it is an opportunity to showcase their facility through the media in a positive light, a desirable outcome when so much in the press about incarceration tends to be negative.

Rikers Island: Finding One's Place in the Garden

BY 11:00 AM the women return their tools to the greenhouse shed and head towards the gate where Officer Ross pats them down before they enter the women's jail. No food from the greenhouse, no flowers, no herbs, no tools, no shards of glass, no decorative feathers, nothing is allowed to leave the garden for the inside. A captain in a housing unit once found a student with fennel seeds she had brought from the garden to make tea. Word got back to Officer Ross who promptly fired the student.

"Nothing, ladies, is allowed inside the jail from the greenhouse," Officer Ross repeats over again like a mantra. "Whatever's in the garden, stays there. If it's on you, its contraband."

The women file out and move down the fenced-in corridor leading to the jail. Outside the fence line, the men, bussed to the greenhouse from the sentenced men's jail across the island, are waiting. Officer Pereira, the correction officer in charge gives them strict orders when they first come to the greenhouse:

"You will see female inmates," he tells them. "Do not, and I repeat, do not talk to any women other than civilian staff or officers. Do not approach the fence with gifts, notes or cigarettes. They will shout at you, yell at you, and maybe even flash you. Do not respond by voice or in kind or you're fired and I'll have to write you up."

Officer Pereira, some 15 years removed from a stint in the army, delivers rules in the staccato command of the military.

He points to the fence. "No one should get closer than ten feet to the fence," he says. "There are remote sensors that will activate if there's motion near the fenceline. Should you have a problem with either staff or inmates talk to me first. Whatever the director tells you to do, do it."

This is their introduction to the greenhouse, a no frills, proceed-at-your-own-risk orientation to gardening. Despite their initial misgivings about any program associated with jail, there is a perceptible change about the men as they pass through the gate and enter the gardens. Their bodies seem to relax as they take in the open expanse of gardens and structures, a physical reprieve form the daily borders of cells, gates, beds and yards.

It is this entry that brings to mind Canada Blackie, a notorious inmate who, over a century ago was sentenced to life in prison for a botched robbery that killed a bank guard. Placed in Clinton Facility, a maximum-security prison in a remote corner of upstate New York, Blackie was a model prisoner for his first seven years. His "good time" came to an end when he shot a guard with a makeshift gun in an attempt to escape. He spent the next two years in the "dark cell," sleeping on a stone floor without a bed or blankets. When Blackie left the "hole," he had tuberculosis and was blind in one eye. Shortly after, he was caught with contraband material in another escape attempt, this time dynamite that had been smuggled into the prison.

Blackie was transferred to Auburn Prison and placed in solitary confinement. During a tour by the new warden, Thomas Mott Osborne, Blackie had a life-defining moment. Osborne was known for his reform agenda, having once incarcerated himself at Auburn to learn about prison conditions firsthand. As he stopped by Blackie's cell, the prisoner inexplicably handed him a makeshift key. Osborne slipped the key into the lock and the cell door opened. As a reward for Blackie's attempt to reform himself, Osborne allowed him to walk from solitary confinement to the main yard, a trip which Blackie later recalled in his journal:

> *"On rounding the end of the cloth-shop, we came into full view of the most wonderful, as well as beautiful, sight I have ever seen in prison—or outside, either, for that matter.*
>
> *I hardly know how to describe this sight: but picture to yourself, if you can, fourteen hundred men turned loose in a beautiful park. These men who I now see among the beautiful flower-beds, instead of the prison pallor and haunted look which once predominated, I now notice smiling eyes, and the clean look which exhilarating exercise in the pure air always brings to the face."*

As a reformer Osborne knew that most inmates would eventually be released, and ex-offenders, broken, bitter and damaged in health, were not likely to transform themselves and become productive and law-abiding members of society. "The system," he said, "brutalizes the men and the keepers. Prisoners are treated now like wild animals and are kept in cages."

Osborne sought to overhaul the system by introducing education, work and religion to the prison system. Inmates were also given time for outside recreation, more nourishing food and less arduous work conditions. He also transformed the bleak prison grounds at Auburn by allowing inmates to build gardens in the prisoners' yard. This humanizing touch helped quell the anger and often-violent behavior that took place when inmates milled about the large Auburn yard.

While a century of reforms have improved conditions dramatically at facilities across the country, the standard time for outside recreation at Rikers is a mere hour per day. The expe-

Daisy, one of two Peking ducks abandoned and sent to us by the Staten Island Zoo, takes a casual stroll by the gazebo. The ornamental grasses provide cover for the pond that flows from a waterfall under the gazebo.

rience of Canada Blackie is not far removed; both men and women entering the Rikers gardens respond in a similar manner. They let their guard down. They smile. Their talk becomes lighter. After one month they enter the classroom with the happy demeanor of a sailor arriving at port after a long arduous journey. Often they are carrying large boxes of milk, Frosted Flakes, bananas, coffee, sugar, frozen hot dogs or hamburger meat—whatever Officer Pereira can scrounge from the mess hall and bring to the greenhouse. Otherwise, the men would eat at their housing complex in the jail, a task that could take an hour off their allotted time in the garden.

Inside the greenhouse I give out assignments. Hariberto M., a personable 35 year-old Hispanic man with years of heroin and methadone use that has left him with one pointed yellow tooth that juts from his palate like a stalactite, is in charge of the kitchen. He scrounges through the cupboard and emerges with five onions and a small sack of potatoes. Hariberto has been in and out of jail most of his life. Along the way he has picked up a number of trades, cooking and carpentry among them. He was also a printer, running a small business that made business cards and brochures. With a good set of teeth he'd be a perfect candidate for finding a job and making a life outside of jail. Only two things keep holding him back: a serious addiction to drugs and his level of schooling, Hariberto is illiterate.

"Don't get me wrong. I like work," Hariberto tells me. "I like getting up in the morning and having a job. The problem is, at some point, I like drugs better."

As Hariberto prepares lunch, GreenTeam Director John Cannizzo interviews two new students and adds their information to our greenhouse database. The database is extremely important, for it gives baseline information on all our students as they enter the program, including their release date, their level of education, their history of employment, contact address and number, as well as their career interests. Shortly before their release we work with each inmate to ensure they have the information and contacts they need to pursue a life outside of jail. It also gives us a foundation for tracking inmates once they leave jail, and to determine, later on, the effectiveness of our program.

When he's finished, John takes the men outside where we keep our carpentry tools, including a large table saw and work tables. He is designing large outdoor planter boxes for PS 96 in East Harlem. With a $5,000 grant, we were chosen to build a small learning garden on part of their rooftop. This is the type of project we especially seek; not only does the grant cover the costs of material and equipment, but it allows our students to make a direct contribution to a community outside Rikers. La Zona Verde, as the project is called, demonstrates how a city jail can be a resource for neighborhoods that are victimized by crime and drugs, often by the very students working at the greenhouse.

John always makes it a point to inform his students that children will be using the pieces they produce so they have to be well made. "Think like you're building something for your own kid," he says. "Build it with love."

John is a good teacher. He illustrates all the steps of construction as if he were designing a how-to manual. He demonstrates carefully how to use the saw and tools, how to measure lines with a tri-square, and the different pieces drawn to scale on the wood they will need for each box.

"It's like a kit," he said. "When we're ready we'll take all the wood to school and have the kids there re-assemble it into planters and benches—maybe a trellis."

The rest of the men are engaged in sets of different projects. Typically, in cold weather we use the time constructing products that will find their way to gardens all over the city. Barkley B., a bricklayer from Staten Island, is putting the finishing touches on a birdhouse he designed and built for purple martins. These migratory birds—the largest member of the swallow family—nest in colonies close to human habitats where they feed on twilight insects like mosquitoes and gnats. Only one colony is found in New York City where, in a salt marsh in Staten Island next to a small marina, a dozen split-level dwellings resting on metal poles rise from an overgrown patch of mugwort. I was told by Gloria Deppe, the retired school teacher who looks after the colony, that the construction of additional houses could increase the nesting population. In light of fears about the mosquito carrying West Nile virus, this is a good thing. I enlisted Barkley to build a prototype house using scraps of expensive hardwood that were donated to the greenhouse by the Steinway Piano factory across the bridge in Queens.

The first house Barkley built was from plans I downloaded from a Minnesota Fish and Wildlife Service website. During

Spring flowering irises (*Iris sp.*) and the deep red foliage of a smoke tree (*Cotinus coggygria*) frame a bird feeder in the bird and butterfly garden.

construction, Barkley and I had long talks on the habits and needs of purple martins: how they no longer nested in the hollow cavities of trees but needed houses constructed by humans; how they liked to be on open lawns near human habitation; how the chicks fledged from the nests; the required size of the entrance (2¼ inch diameter) and its proximity from the base of the platform (1 inch).

"I think about these birds all the time now," he tells me. " I spent half the night wondering what would happen if you painted the house a different color than white (martins nest in white structures). Would they move? Sue the contractor for breach of contract? I mean, this is New York. We're talkin' free housing!"

The first house took the lines of a classic purple martin house, two-stories, 12 nest holes, white paint and a pitched gray shingle roof. All it needed was a weather vane and it could have been a feature in a Lands End catalog. I drove out to Lemon Creek Park and proudly displayed Barkley's house to Gloria Deppe. She promptly rejected it.

"There're no sliding drawers," she said. "The sparrows and starlings get in there and lay their eggs before the martins. Someone has to check each nest box and destroy the eggs. Unless of course they're martins. The nest box has to slide in order to do this."

Barkley was not happy to hear this. His first impulse was to tell me to take the house and shove it. I understood the sentiment. Barkley was short-tempered. He was serving a year for exacting his own brand of retributive justice with a baseball bat after a drug addict had broken into his house and removed some of its contents. He had spent many of his younger years in and out of youth homes and detention centers. He had a lot of anger and jail was not making him any less angry. But I also sensed a dramatic change in him after some months at the greenhouse. From angry and taciturn, Barkley began to open up. He talked about his background, his inability to control his emotions and his problems with ex-wives, all of which led him back to jail on more than one occasion. The garden had a way of disarming, of suspending judgment, of giving him the space and time to reflect on the past and think about the future. Could he do things differently?

After considering the required changes he grabbed a pencil and began designing a new structure.

The second house was one-story with a pitched roof that had a perching bar stretched across it. There were security rails in front of each nesting box with a small balcony so the chicks, before they fledged, could safely climb through the hole and hang out without falling off the edge. But Barkley had miscalculated his measurements. The platform was too short and the security rails prevented the nest boxes from sliding all the way out. An individual could easily check the nest, but could never remove an egg. This one Barkley rejected on his own.

The third one he built was slightly modified with a larger platform and a hexagonal shape and drawers that could slide all the way out. "I want to get this right," he told me. "It's good for me to focus on something positive. I mean, let's face it. If it wasn't for coming out here I'd be inside beating someone up."

As spring approached Barkley was anxious to finish the house. He applied two coats of white paint and, when it dried, fastened a dowel to the length of the roof that the martins would perch on. He knew that by the end of February the first birds would be winging their way up the Atlantic seaboard from Brazil. They would follow the insects, gradually moving north until arriving in Staten Island by May. The first birds act as scouts, searching for and staking out suitable housing for the slower moving colony. And there, in view of the Manhattan skyline, the martins would see from above a small cluster of funny little houses poking from a salt marsh. There they would spend the summer nesting, raising their families and eating the insects by twilight.

Barkley was ideal for this project. Building a birdhouse at Rikers was more meaningful to him then simply helping the martins survive in the city. I could sense the change in him as the third attempt at constructing a suitable dwelling progressed. He seemed less on edge, less defensive in his surroundings, particularly when the officers, in sport, would ride him, trying to get a rise. He took pride in the house, often stating how irresistible the martins would find his craftsmanship. "Choice housing," he would say. Besides, Barkley was from Staten Island and knew the park where the birds liked to nest. The birdhouse was his contribution to society and it

Growing Trees in Jail: Santa Rosa, California

At first glance, the fields adjacent to the North County Detention Facility in Santa Rosa, California, seem to be sprouting branches. A closer look reveals rows and rows of container-size trees that will one day supply San Francisco and other Northern California towns with street trees. The program, a unique partnership between the Sheriff's Department, Sonoma County ReLeaf, and the National Tree Trust, puts male inmates to work growing and nurturing over twenty different varieties of trees from seedlings to full planting size. The trees are then given, free of charge, to schools, county municipal agencies, and non-profit groups as long as they are planted in parks, schools or other public spaces.

The program began when Rick Stern, an agriculture instructor at the facility, decided to establish a "tree growing-out station" in Sonoma County. He contacted the National Tree Trust, which agreed to supply trees and funding, while Sonoma ReLeaf, a local non-profit signed on to help coordinate the program and provide financial management and tree planting services.

In 1996 the program was officially underway with the arrival of the first batch of seedlings. Today that number has grown considerably, sometimes reaching 15,000 trees available. Stern is a believer in variety and grows coastal redwoods, valley oaks, maples, redbud, sequoia, and others.

According to Stern, everybody wins with the program. "The county gets a source of free, healthy trees, which benefits the environment," says Stern. "And it gives the inmates a chance to learn responsibility, good job skills and the belief they can contribute to society while they're serving time." ■

COURTESY SONOMA RELEAF AND NORTH COUNTY DETENTION FACILITY OF SANTA ROSA.

helped him connect to his community, making his stay at Rikers and his arrival home that much easier.

But Barkley would not articulate it in that manner. Instead, he would say that after his release he was going to take his daughter to Lemon Creek Park and show her the martins and the house that he built for them. He'd repeat it, like a mantra, until the day he left Rikers.

CHAPTER FOUR

Starting and Expanding a Horticultural Program

U.S. prisons today offer fewer services than they did when inmate problems were less severe . . . just one-third of all prisoners released will have received vocational or educational training while in prison despite serious deficiencies in these areas.

—JOAN PETERSILIA, AUTHOR OF "WHEN PRISONERS COME HOME"

In the eight years since the GreenHouse jail-to-street program was established on Rikers, I am often asked how to start similar projects elsewhere, in prison facilities, with at-risk youth groups or with institutions working with the mentally ill. Besides being one of the few lines of work that has therapeutic qualities, horticulture offers unlimited opportunities for trained and untrained labor. The Department of Labor has listed horticulture and landscaping as one of the growth industries of the 21st Century with over $1.5 billion dollars of business generated in 2003. Our Rikers program taps into these opportunities with our GreenTeam, which students can join after their release from jail as paid interns, earning $7.50 to $10 an hour on different gardening projects around the city. After a year of training, the interns work on obtaining more advanced jobs that can lead to long-term careers. (see Chapter Nine)

Programs such as ours impact more than just individuals—much of our work focuses on habitat enhancement and environmental restoration on public and private lands. Our work also serves community groups and institutions. For example, during the summer the interns give workshops to children on building butterfly gardens at New York City housing projects. They work in a psychiatric hospital with patients who have an established vegetable garden and in a green market that provides local residents in a low-income neighborhood with freshly grown produce.

For state and federal prisons or county and city jails, opportunities are plentiful for developing a comprehensive horticulture and gardening program. While funding for programs throughout the nation have been curtailed or eliminated, prison administrators are, in general, supportive of rehabilitation and will likely welcome projects that do not rely on money allocated from the facility's budget.

Teamwork is essential in the prison environment. The warden, or the deputy warden of programs and security, make the day-to-day management decisions and must be committed to the project from the very beginning. Demonstrating that a program is cost-effective and produces services for the jail, distinct from its benefits to inmates, increases the chance for a project's approval. For example, plants grown in a facility greenhouse or provided through donations can be used to beautify the facility grounds, and food and herbs can supplement the facility's food budget.

While start-up costs seem negligible, the facility must provide officers to guard the students, secure transportation, and address potential security issues. This can offset or dampen a warden's initial enthusiasm. A proposal, detailing benefits and costs, with potential funding sources is instrumental in securing the department's interest in backing the program. In addition, be ready with a wish list of supplies in case the department's annual budget leaves the warden feeling generous and supportive of the program.

A Word on Lesson Plan Activities

Our experience has found that inmates of all educational levels readily embrace activities that teach scientific processes in botany and horticulture. These activities can include soil testing, experiments with phototropism and transpiration, flower dissection, and seed germination. Lessons can take place on days of inclement weather or prior to beginning the day's work schedule. You may be surprised at the interest inmates take in their work based on their increased understanding of nature. ■

Many prison facilities already have small gardening or greenhouse programs. These can be expanded to include landscape details, construction classes, nurseries, herb gardens and plant propagation or turf management programs with a focus on vocational training, science studies, and job skill development.

When devising a program for a jail or prison, it is necessary to adapt it to the social and environmental conditions of the institution. The regional topography and climate as well as the availability of land or greenhouse will determine what can be grown and when. Although the administration has the final word the extent and manner of participation and the kind of work allowed in the facility, the program director creates the vision for what is possible. As one goes about starting or expanding a horticulture program an assessment of the site is essential and the following areas of consideration should be examined:

Assessing the Physical Resources

Assess cultivable land

A small plot the size of a kitchen garden is sufficient to run a hands-on learning garden that could support up to 10 inmates. Of course, the larger the land the more opportunities there are to establish an integrated system with food and herb production, landscaping, plant production, water features (ponds or streams) and if possible animal husbandry (chickens, ducks or livestock). It is necessary to work closely with the warden in identifying potential areas around the facility for beautification, wildlife habitat and cultivation. Sightlines are critical in a correction facility, therefore, what may seem like an expanse of empty land is a security feature of the facility. In some cases vegetable gardening or small annual plantings are more acceptable to the facility than a landscape of trees and shrubs that obscures the sightlines.

Establish a greenhouse

Greenhouses provide the opportunity to grow and propagate plants year round, and boost production for in-house or community use. It also centers the program, creating a space where participants meet, plan activities and conduct lectures. A skilled crew can build a fairly inexpensive structure, or a commercial nursery that may be willing to donate one it is no longer using.

Make room for a nursery

Many commercial nurseries will donate stock that is damaged, undersized or difficult to sell. Creating an area to grow donated shrubs, trees and perennials that can be replanted around the facility or in the community is a major benefit for the program. At Rikers, we receive annually well over $15,000 worth of plant material that is distributed and planted in different community settings throughout the city.

Secure a classroom or shelter

An area indoors to conduct classes and provide shelter on rainy, cold or hot days is crucial for a program. In negotiating with the administration, always inquire about the availability of a secure space that can be used for lectures or workshops.

Develop space for a construction workshop

Inquire about the possibility of developing a carpentry class for building cold frames, composting bins, pergolas, window and planting boxes, bird feeders and other related outdoor items. This adds not only a new vocational avenue but also provides numerous activities for winter, as well as useful materials for off-site gardens. At Rikers, students have built a number of rooftop gardens (reassembled on-site) or planter boxes for public schools or non-profit organizations, as well as bird and bat houses for public parks and libraries.

Find sources for free mulch and compost

Tree care companies often are willing to donate woodchips for mulch, and townships are increasingly constructing organic compost facilities to reduce green refuse waste. Rural farms sometimes have a surplus of slurry (processed manure) or manure they can donate for compost or fertilizer. Rikers has a state-of-the-art compost facility built by the Department of Sanitation that provides compost for the Rikers gardens and for projects off the island.

Guidelines for Managing a Successful Program

Be realistic

Setting realistic goals for both the students and the staff creates a framework for success that builds on and sustains itself through the life of the program. For example, students may have a learning disability, history of drug use or trauma-related mental disorder that has kept them at a level of low literacy and unemployment most of their lives. Although the program goal is to provide inmates with career jobs in horticulture it is unlikely that within a year students will be transformed into professional horticulturalists (though that possibility is certainly never ruled out). Rather, more moderate goals should be established: in five months students should be able to grasp fundamental basics of plant science or landscape design; they should be able to execute proper techniques in plant care and garden installation; and they should have an understanding of plant, animal, and human relationships.

Work goals should be reasonably set so that students and staff can feel a measure of accomplishment in the program. Many students have a history of failures in their life, and if they fail to meet either personal or project goals, they easily become frustrated and demoralized. Completing a series of small projects builds self-esteem and a foundation of success and achievement.

Create a work environment that promotes program goals

It is essential that the work environment be designed to meet the needs of both the staff and the students being served. If therapy is a major goal of the program, the instructors and correction officers must work with the inmates in a way that encourages communication, trust and respect. This openness is important when horticulture is being used to address issues affecting the students' behavior. If the focus is on job development and job readiness, the environment should mimic the expectations and conditions of the workplace.

The Rikers GreenHouse, with its conjoined classroom, greenhouse and gardens, gives the instructor considerable freedom in establishing an integrated program that reflects the diverse needs of its students. For example, the large gardens surrounding the greenhouse are conducive to private conversations that allow the instructor to work one-on-one with inmates in relative privacy from other students or officers. Simultaneously, lectures can take place in the classroom, a meal using fresh food from the garden can be prepared in the classroom kitchen, and a professional pruning workshop can be conducted in another section of the garden.

A trellis built by students makes an elegant statement at the entrance of the classroom.

Horticulture programs should be equipped with adequate tools for the students to pursue their day's work efficiently. Apart from gardening or construction materials, resources that focus on particular subjects such as Integrated Pest Management, ethnobotany, or herbal remedies. or computers for professional development, books for a library and any other resources that strengthen the student's ability to learn and ultimately function in a job outside of jail are essential for an effective program.

Choose an instructor

Whether your instructors are drawn from the staff or from the outside, they can transfer core knowledge through teaching and gardening programs, but without leadership, important goals cannot be accomplished. Instructors must tread a fine line between serving the needs of the students and strictly adhering to the regulations of the facility. It is not enough to simply establish relationships with the students. It is equally important to forge strong bonds with the supervising officers.

The instructor must be flexible and aware that enforced instruction is not always effective. Informal work sessions provide inmates the opportunity to explore their work environment and experience a reprieve from the regimentation of life in prison. A balance between formal and informal training and work projects is crucial. The instructor should be able to tap an individual's motivation to learn skills critical to finding meaningful work on the outside.

Use a team approach

At the work site, a meeting should be scheduled every day to discuss job activities of the prior day and plan the tasks for each day. These regularly scheduled meetings help sharpen the focus of the crew's efforts and track how well the curriculum is being implemented. Regularly scheduled meetings give the crew the opportunity to develop cohesiveness as a group, identify problems affecting the success of work being undertaken in the garden, and to receive positive reinforcement for successful efforts.

Give students clear instructions

Each student enrolled in a program could be asked to sign a contract that explicitly describes his or her responsibilities and the responsibilities of the staff. Such a contract may require the participant to follow the instructions of the staff and to be on time and ready to work. The students may be required to actively take part in self-assessment and maintain contact with the program after their release. Such agreements can make it possible for instructors to contact former students for a prospective interview should a job offer match the students' skills.

Establish pre-release programming

Information about support services should be available for students as they get ready to leave jail. What are their needs? There are numerous substance abuse programs, health clinics, legal aid, subsidized/Section-eight housing and shelters, and vocational training programs that ex-offenders are eligible for (*see Aftercare Chapter*). Instructors should help students make career plans by being informed about the different opportunities in horticulture and its related fields. Interviews should be scheduled with prospective employers whenever possible, and problems that are barriers to employment should be addressed.

Develop a network

Because most programs operate with limited resources, garden or greenhouse programs may need to depend heavily on a network of private and corporate funders, concerned individuals and businesses, and governmental and non-profit agencies. A strong network improves the services provided to the students as well as services that can be provided to the community. It also increases public awareness of the program mission. Contacts on the outside can offer important resources, work opportunities and funding revenues. Nurseries can donate plant materials or seeds. Establishing relationships with nurseries, schools, and accredited associations and foundations are essential to a successful cost-effective program.

Be creative and draw on all available resources

Horticulture is wide open to a variety of work-related skills training. Computers can be used for different types of landscape programs, landscape design or for creating in-house newsletters. Inmates with artistic talent or construction or trade skills can utilize their knowledge for different gardening related projects. Outside experts in botany, landscape design, writing, art, tree maintenance or construction can also be brought in to provide workshops.

Understanding the Inmate Population

There is a dramatic difference between working with sentenced inmates in a prison facility and detainees in a jail. Prison, on the state or federal level, is for individuals sentenced to do time. Jail holds primarily detainees—individuals who are waiting for trial or court dates and have either been denied bail or are unable to pay it. Rikers contains both sentenced inmates—anyone sentenced to a year or less stays on Rikers—and detainees.

Sentenced inmates offer a more stable population to work with, while detainees tend to be transitory. There are also differences to recognize among age and gender groups. Each program should be tailored to the specific characteristics of the inmate population. The following offers some general guidelines:

Long-term sentenced inmates provide the opportunity to create an effective programmatic hands-on course that leads to certification by the American Nurseryman's Association. Inmates involved for at least two years and longer can see through the design, construction and maintenance of gardens and landscape systems on a consistent basis. Such programs can incorporate a strong mix of classroom instruction with horticulture fieldwork. With a stable population, the program can include more advanced aspects of horticultural training,

including landscape design, computer graphics, propagation and sophisticated pruning such as espalier and fruit tree cultivation. Planning is long-term and can lead to highly accomplished group or individual projects.

Mix of annuals, grasses and perennials as spring arrives in the bird and butterfly garden.

Short-term sentenced inmates require a program that allows them to experience different aspects of horticulture without requiring too much classroom learning or certificate courses. Many short-term sentenced inmates may only have one or two seasons of horticultural work, and will not be expected to leave the facility as accomplished horticulturists. More emphasis should be placed on creative projects and basic vocational skills that can lead to further interest in training programs or internships after their release.

The Massachusetts Story: North Central Correctional Institute

by Greg Barnett, Coordinator for Gardener Correctional Institute

The Gardener Correctional Institute, in Gardener, Massachusetts, has a long-running program that allows teams of five or six inmate gardeners to tend individual garden plots. In a facility where inmates may be incarcerated for long-term sentences, obtaining a garden plot is a special privilege. A large number of Gardener inmates are serving life and double-life sentences, while 50 to 60 percent of the population are incarcerated for sex offenses.

It's easy to see why individual plots are so valuable to inmates, who are allowed to keep or barter any produce they harvest. The most experienced gardeners adopt a leadership position, handing out weeding and watering duties to other inmates. As the harvest draws near, newer members are assigned to guard plots. The year after year cultivation of their vegetable beds provides a stable framework for inmates serving long-term sentences. Vacancies become available only when inmates are released or transferred.

The garden plots supplement the prison fare with fresh produce. The gardeners compete with each other to see who can grow the best tasting or largest vegetables. The gardens promote self-reliance within a setting of cooperation as expertise is passed from one generation of inmates to the next—not unlike the workings of a small family farm.

Many offenders who never voted, held a position of responsibility or even expressed an opinion on any community matters find themselves practicing a form of democracy based on "private property" and shared responsibility. Prisoners are able to pool resources while each is free to use his share any way he chooses. Some prefer to simply consume their produce, while others use it to purchase services and goods from other inmates in a kind of micro-economy. These garden plots become mechanisms for problem solving, models for rehabilitation programs, opportunities to teach management concepts, tools to illustrate the importance of planned rather than impulsive behavior, and channels for inmates to translate their personal experiences into a community effort. ■

Detainees are the most transient. Since they will come and go with sudden releases, court dates or transfers to state or federal facilities, they create a relatively unstable work force. Work activities should center around on-going projects with an emphasis on therapy and basic skill learning.

Adolescents (16 to 18) generally require both structure and a hands-on work. Many adolescents are still working towards their High School Diploma or GED (Graduate Equivalency Diploma). A course with emphasis on science is effective for developing links to educational programs at the facility's school. A hands-on educational horticulture program is especially effective as an outlet for restive adolescent energy and a motivational force for school learning. Projects should be short-term, creative and offer a high level of visible accomplishment.

Older Populations require less monitoring and can be assigned long-term individual tasks and projects. Many of the older inmates have already established job skills. Gardening offers a creative way to release stress, work out individual problems, and generate interest in horticulture as a long-term hobby. Sometimes individuals are highly skilled gardeners and can offer instruction to newer or less experienced members of the program.

Mixed Populations primarily occur in county or city jails. In some instances, such as the women's program at Rikers Island, the population is extremely mixed with short and long-term detainees, short-term sentenced, older and adolescent women together in one program. This demands a program with a large degree of flexibility in both goals and objectives.

At Rikers, experienced gardeners or longer-term program members are given specific assignments or projects, and train new students assigned to their particular station. Short lectures, for example, in soil, pruning, plant science or design take place several times a week. Tasks change based on demand and season. Vocational training is offered to those individuals expressing a direct interest in horticulture as a potential career. If a student is having a troubling day, the instructor might assign an interesting work task to alleviate stress. Creating topiary, planting a tree, building a rock garden and harvesting herbs are selective tasks that students especially enjoy. Meanwhile, Rikers instructors work at finding programs for released adolescents (ages 18 to 21) that help them finish High School or their GED, enroll in college, or find job training or jobs after their release.

The creativity inherent in horticulture and gardening is especially conducive to melding disparate populations into effective work groups while creating space for individual needs.

CHAPTER FIVE

Design: Transforming The Prison Landscape

Prisons should be a tour through the circles of hell where inmates should learn only the joys of busting rocks.

—FORMER MASSACHUSSETTS GOVERNOR WILLIAM F. WELD

Rikers Island: Garden of the Guards

THE SUN AT NOON shakes the chill of morning as the men break ground on the last phase of a major project: landscaping the gazebo and pond, which were built the previous fall. Unlike traditional gazebos that are built ground-level, this one is raised almost four feet with steps leading to the flooring. A stone built waterfall flows from beneath it and falls into a small pond that by late spring will be stocked with fish, frogs and turtles. The culmination of group desire, multiple designs, salvaged wood, and the skills of a talented, incarcerated carpenter, the gazebo is the structural and spiritual centerpiece of the garden. More satisfying, I had very little to do with its planning or construction. I approved the designs, but everything else was between Officer Pereira and the inmates.

Periera, who on most days seems disengaged to the point of absence—I imagine he could watch Rikers and everything on it sink into Flushing Bay and not blink unless his pension sank with it—has a weakness which belies his seeming indifference. He's a compulsive problem solver. Electrical wiring, plumbing, carpentry, computers and maintenance are sources of an endless stream of questions; all I needed was the right combination of technical glitches and I could capture his interest. Without Periera's involvement a project this size simply couldn't happen.

How successful we were in the garden depended in large part on the officers assigned to the program and the manner in which they treated the participants—as inmates or students.

"When you work inside the jail," Pereira once told me, "It's easy to forget that the inmates are individuals. You just treat them all as inmates—the good, the bad, and the pain-in-the-ass. But in the garden, it's different. I really want to see these guys succeed.

"I'll never work inside the jail," he added. "I'll take early retirement before I go inside again."

Who could blame him? Pereira came to Rikers in the early 1980's when, in one officer's words, "the jails were run by inmates." Violence, gangs, drugs and weapons were rampant features of everyday life behind bars. Most officers simply wanted to get through their shifts without incident and get home to their families. By 2000, newly created units such as Gang Intelligence, a beefed up corrections force, and changes in the ability of the Department of Correction to prosecute crimes that occurred behind bars had made a sizeable impact on violent incidents at Rikers.

Despite that, the stress of working inside jail, the hatred, the anger, the foul language and the lack of visual comforts take a constant toll on human spirit. No officer is completely immune. The garden was therefore, as much a sanctuary for Periera as it was for the inmates. And it showed in his relationship with the students.

John P. arrived at the greenhouse in the fall of 2000. A wiry, grizzled man in his late 40's whose life alternated between his time spent in jail and his work as a carpenter, John took one look at the gardens and quickly suggested building a gazebo. I readily agreed and suggested placing it near an existing seven-by-three foot concrete hole that I thought would make an attractive formal pool. John came up with a design that showed a waterfall flowing from the gazebo into the pool, which on paper we quickly enlarged into a pond. I gave the plans to Officer Pereira. What could we use as corner posts to hang the platform high enough for a waterfall to flow beneath it? What could we use to construct the waterfall? How could we circulate the water? What kind of filtering system would keep the pond clean? Periera looked at the plans and gave a non-committal, "Let's see."

John the carpenter applied his skills building the gazebo/waterfall/pond complex, interior cabinets, a shed; helped with wiring and plumbing, ran literacy classes, helped in the law library and painted murals. It was outside jail where life skills failed him.

The following day, as the women left, the men arrived and began unloading sixteen foot-long eight-by-eight inch posts along with 60-pound bags of cement that we foraged on the island. The project was underway. John measured out the post-holes and with two men began the task of digging holes and setting the corner posts parallel to the pond site. Over the next week, Periera scoured the island on a scavenger hunt for planks, two-by-fours, and anything else that could be used for building the gazebo. As the wood piled up we took to the Rikers shoreline for rocks. Pereira would stand at the crest of an embankment as inmates scrambled up and down hauling

Above and facing page: The gazebo being constructed with salvaged wood and later with the path in place, fully landscaped.

rocks large enough to serve as a base for the waterfall. When the boat patrolling the island rounded the bend, Periera would order the inmates onto the bank away from the water.

Periera told them, “If security sees us and stops, I’m gonna tell them you were trying to escape.” The men, not sure whether to believe him or not, laughed nervously. “I got a pension to protect,” he added.

Off the island, Periera made pit stops at Home Depot for deck screws, drill bits for his own drill, blades for his circular saw and gathered materials he had lying around his house in Staten Island. By the end of a month the gazebo was almost finished, complete with a lattice foundation, which John P. stripped from two-by-fours into thin pieces that he nailed to a frame. The lattice covered three open sides beneath the raised floor and a set of stairs; the fourth side would be the stone waterfall. With a router, John grooved a railing for the platform and banisters for the steps. During breaks Pereira would ask him about flooring and plumbing, since he was planning to rebuild his bathroom.

“I didn’t know anything about this stuff, until I bought a house,” he told me. “I had to teach myself because I couldn’t afford anyone to do it for me.” Officer Periera grew up poor. His father, a drug dealer, was killed on Grand Street and for the

most part he was estranged from his mother and his two brothers and a sister. He got his first taste of the outdoors as a kid milking cows on a farm upstate through the Fresh Air Fund—a program that provides inner-city kids with several weeks of summer vacation in the country. He joined the army and after three years driving a tank in Germany, he settled back in Brooklyn with a wife and three sons. For a short while he rode a horse as Parks police, mostly in Central Park. The city then hired him for Correction, a choice he made strictly because it paid better than other city and state enforcement agencies.

As an officer, Periera made the kind of move known to many of the middle-class, trading an apartment in an inner-city neighborhood for a house in the Staten Island suburbs.

Pereira, along with Officer Ross, spent eight years working

plants trees that provide food and habitat for birds, or perennials that sustain larvae of a specific butterfly, or herbs that were used by a parent for cooking, connects to the environment in a manner which transcends the tasks involved in grounds maintenance or beautification. The act of planting not only becomes a personal experience, but the decision of *what to plant* provides an avenue of self-expression and empowerment. The process results in a prison environment that is more conducive to the management and care of offenders, and a less stressful workspace for the staff and officers.

Imagine bare space transformed into vegetable plots with circular herb and perennial gardens that soften the contours of the landscape. Small trees interspersed throughout the facility with an understory of bulbs and perennials provide increasing diversity and three-season interest. Walkways through gardens of native trees and shrubs bearing fruit attract bird life, while selected perennials and annuals draw butterflies. The gradual development of bleak ground into a dynamic landscape is a series of steps that humanize facilities and calm the aggression and anger which are part and parcel of the prison system. The process begins with design.

The Creative Design Process

The process of designing gardens in prison is rooted in both a mental and physical space where students integrate a working knowledge of plants, design principles and their own creativity to alter the confines of their physical surroundings. The metaphor of design is poignant: designing pathways, establishing boundaries and entranceways, choosing the right plant for the right place, and using what nature provides (stone, wood, water) to embellish the garden takes on the psychological contours of the student's own needs and desires.

In starting a new garden project at Rikers, students typi-

The Gardens at Rikers

An overview of the gardens and fence line in the two-acre greenhouse facility.

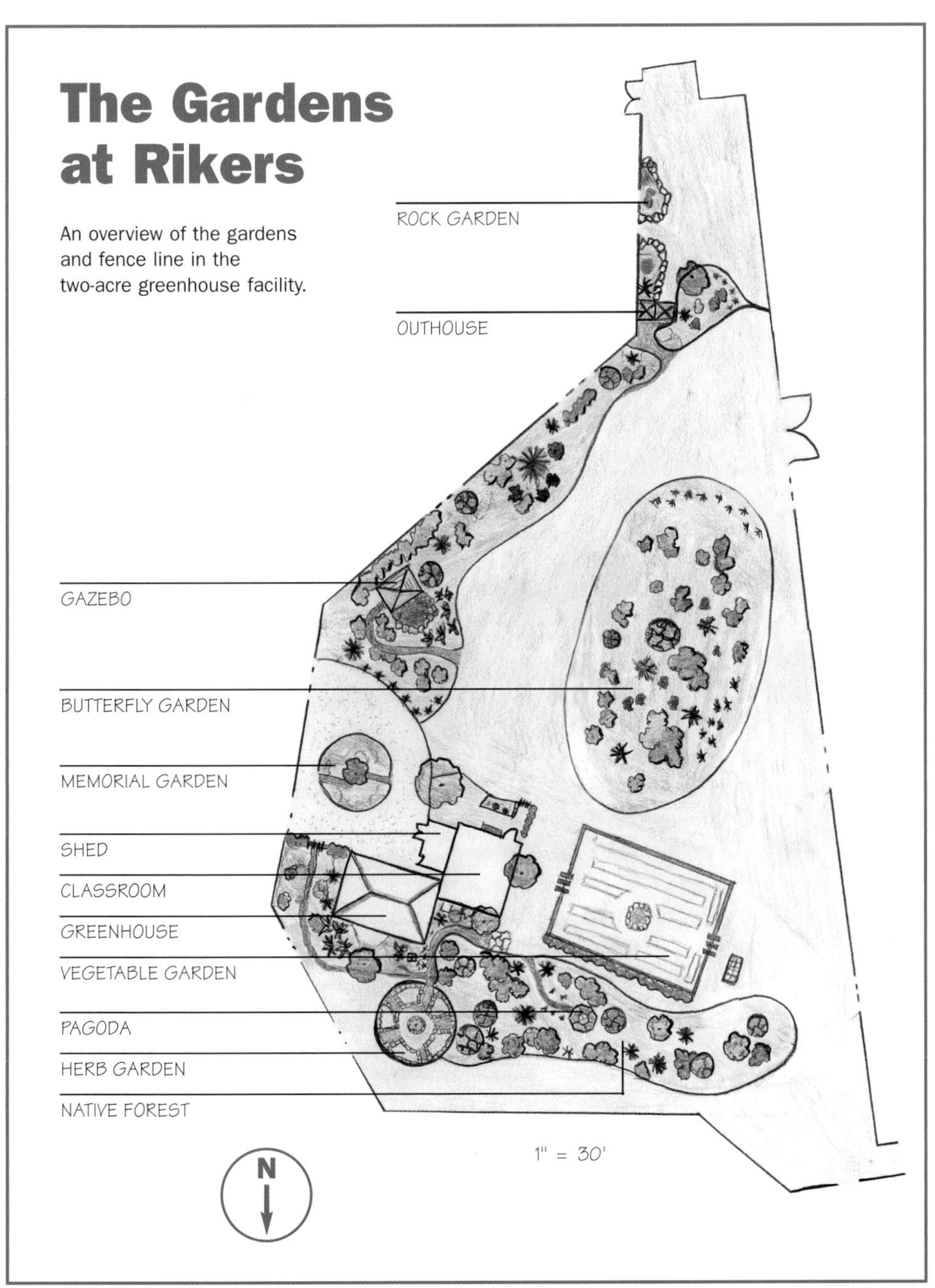

The Process of Design: The Prison Experience

By Paula Hayes

The usual cultural padding we associate with institutions or communities and neighborhoods does not exist in the prison landscape. Nothing seems to be decorated or celebrated. Driving onto and across Rikers Island casts an immediate feeling of the physical reality that there is no place to hide. Home is, in a way, a place to hide and our designed landscapes are about providing ourselves with places; places to be alone or to be alone with a particular person or group. The absolute erasure of this possibility—to go somewhere, to hide, to be in a personal space—pervades the prison environment.

On Rikers, within this relentlessly exposed landscape, exists a radically different approach to our perceived sense of incarceration. Deep within the complex of facilities, checkpoints and fencelines is the prison garden and greenhouse where the jail houses its program and where I was invited to teach a day's course in landscape design. Clearly, the greenhouse and grounds were an oasis for the inmates involved. Under the shade of a massive crab apple tree were benches and tools for gardening. It was summer and the tended areas were flowering and attracting birds and bees. Water was flowing; all was nurtured and cultivated.

I entered the greenhouse classroom where a group of women were waiting for me. They seemed tired (many of the women at Rikers are incarcerated for drug use and are medicated or on methadone), but clearly glad to be outside; the door was open to the breeze. I began talking about what I imagined to be the most difficult aspect of their time in jail: the inability to be alone, unwatched, to quietly restore themselves on their own terms. I tried to equate gardening and garden design with the process of self-creation: attending patiently to the subtle movements of growth is the core of providing oneself a private arena to lead an explored life. To be in the landscape—moving through it and using its provisions—enlivens us as caretakers of the earth and also of ourselves.

We talked about design elements for a garden the women were creating in memory of an instructor who had passed away. We drew sketches on the blackboard examining entranceways and exits and pathways through the garden. We talked about boundaries and private spaces; how gardens can create a series of separate rooms that can be either real (using hedges) or illusionary (through borders). We looked for meaning in shape, color, texture and fragrance. We then went outside. The students clearly wanted to move and work. We measured and laid out our plants and pathways according to the sketch. We watered the new plantings. By the time we had finished I felt that having this chance to build the garden was a profound and lasting experience for all of us. ■

The memorial garden designed and planted by the students and designer, Paula Hayes.

Paula Hayes is a New York City-based landscape designer and artist.

cally identify an undeveloped area in the two acres surrounding the greenhouse and ask themselves a series of questions. What are the relevant features of the site? How can the site be integrated into the existing landscape? How does one create beauty and its attendant natural functions? How does it benefit people and wildlife? What is its use? The position and movement of the sun, the site's access to water, its slope, and its soil type are important elements the students consider in shaping the land to its intended purpose.

Form and aesthetics are the next stage of site development. Will the garden or planted area have curves and edging, berms and contours, multiple layers of height, a variety of colors? Will there be complementary structures, access, shade, and symmetry? What is the vision the students hope to create? And, ultimately, who will experience the end result? Officers, staff, visitors and other inmates are all part of the prison environment and their enjoyment of the landscape helps build the self-esteem of the inmates who created it.

With the answers at hand, the students develop their preliminary sketches. They can focus on the future, imagining how the site will develop, the growth and structure of the plants they choose, the color of their flowers and texture of the foliage. Because the need to maintain sight lines are a constant factor at Rikers, tree size and placement is a major concern, and trees are used judiciously in the gardens.

There is considerable direction provided by the instructors in terms of what is both appropriate and doable. A cement walkway, for instance, may not be desirable for a natural woodland garden; nor is a costly bluestone patio viable for a program short on funding. Plant choices are debated and students must defend each choice. There is considerable give and take as well as some frustration. Finally, when the inmates and staff reach a consensus, the sketch is enlarged and drawn to scale. The students use professional symbols to produce a final design. The design's installation will undergo several permutations as donated materials are substituted for materials that are either too expensive or unavailable.

We often say that landscape is the ultimate painting and sculpture, a three-dimensional work of an artist whose technique is based on a working knowledge of science. Every plant has an appropriate place based on its function and adaptation to an ecological niche. Some survive better than others. Some will colonize and exploit a site, eventually spreading out with their roots or taking over in succeeding years through the dispersal of multitudinous seeds. Others are exotic and need excessive care. Trees create micro-climates and habitats that can alter a site dramatically and create new opportunities for diversity and plant structure. Herbs not only provide fragrance and color, but are used as traditional medicines, salves and food by different cultures worldwide. All of these elements come together in the garden.

As students come to understand the characteristics of different plant species and varieties—their growth rates and size, their specific flowers and fruit, their structure and foliage and their specific niche in the garden—they begin to understand their own development and growth.

The ability to control nature is inherent in gardening, but control is balanced with the need to minimize maintenance and provide opportunities for plants to naturalize and sustain themselves over time. A well-designed garden is a social contract with nature, a balance between art and the environment. This relationship, together with an accumulating bank of knowledge and skill, are the ingredients necessary for bringing about a landscape transformation.

CHAPTER SIX

The Science of Gardening

In nature there are neither rewards nor punishments. There are consequences. —ROBERT INGERSOLL (1833–1899)

In learning about the relationship between gardening and ecology, students learn to do more than build pretty gardens: they learn to build systems that improve and sustain the local soil, plant and animal complex, which in turn sustains terrestrial life. While it is not expected that students become ecologists during their time on Rikers, a few concepts in ecology are instrumental in promoting the idea of "natural systems" gardening. An introduction to this begins with a lecture called "Ecology and the Natural Garden" where I cover the diversity of the earth's ecosystems, including our own urban ecology. The plants and animals of these systems form complex interdependent relationships that secure their ability to adapt and survive in a given environment. How these relationships are managed will determine the students' success in making a positive impact on the system's ecology and, to a larger extent, on the regional ecology.

Rikers Island: People are Nature, Too

ONE DAY Pereira brought a backhoe driven by a contractor who was doing work near the greenhouse. "He owes us a favor," Pereira says. "I gave him some plants for his trailer." The backhoe scooped out the pond which the men had already begun digging with shovels, saving a week of back-breaking work with the push and tug of a few levers. Sand was poured and a rubber lining was stretched across the bottom.

By Christmas, the stones for the waterfall were in place. The foundation for the stones was cemented, and the top stones, which directed the water and created the "splash zone," were held in place by gravity and weight. Three holes spanning the top of the waterfall were left for hoses that were attached to the main hose using a multiple outlet. This way, the water would descend in three streams that would eventually meet at an overhanging rock near the bottom making a formidable splash into the pond. But cold weather froze the hoses and the waterfall was shelved for the winter.

By the first week of January, John P., the brilliant carpenter who designed and built our outdoor complex, was released from jail. For two straight days prior to his going home, he was perched on the roof of the gazebo hammering shingles (red for their resemblance to baked tile). On his last day, he hammered in the final piece, stood on the corner of the roof and in the manner of the actor Leonardo De Caprio in the film *Titanic*, yelled, "Look at me. I'm King of the World!"

Tyrone, a young inmate who worked on the project as John's assistant, turned to me and said, "This is the first time I ever actually finished something I started."

This desire to build a three-dimensional structure and, in our case, a structure and garden from plans and drawings is powerful in the context of jail or prison. Typically, an inmate new to the greenhouse will utter something along the lines of "I'm not too thrilled building something for Rikers Island, so don't expect much out of me."

Invariably, I counter with "You're not building it for Rikers, or officers or for City Correction, you're doing it for yourself." And, typically, those words when spoken are meaningless. Yet, after a few weeks, when the inmate spends time outdoors, away from jail, with good food, and with officers who will leave him alone as long as he respects greenhouse rules, work falls into a daily rhythm, and at some point there is a personal transformation. It may occur when an inmate observes his/her role in transforming a cup of dirt into a germinating plant, some scraps of wood into a birdhouse, or a barren piece of ground into a bed of flowers. It may also come with my exhortation that the better the plants are cared for, the more vegetables and herbs we'll have to eat. Or it's simply when an officer or civilian staff takes the time to thank the students for the beautiful and professional job they did building the garden.

Accomplishment, knowledge and skill reinforce each other. The work becomes infused with ownership and pride, and pride with creativity; the product manifests itself as a permanent mark on both the landscape and the person. If I could make one of those inspirational banners that coaches like to hang in high school locker rooms, and hang it in the greenhouse, it would say, "Accomplishment is good. Don't take it for granted." To achieve something and have it remain behind for the next 50 years is downright powerful.

The Garden as a Natural System

Gardening should not only be environmentally sound, but also increase the health and vitality of the regional ecology. Natural systems of gardening depend on the same type of inputs as non-organic gardens, such as fertilizers, new plants, seed dispersal, pest control and watering. The difference is that natural systems are self-supporting. They utilize the processes inherent in nature and eliminate the need for petro-chemical fertilizers, toxic pesticides and practices that are harmful to the environment.

Walk in a forest. That spongy ground underfoot is decaying leaves and twigs that are turning into humus, decaying organic matter that improves fertility in the soil. This spongy material insulates the soil from the cold, keeping plant roots warm during winter. In the summer, the spongy material, known as duff, keeps water from evaporating from the soil, keeps plant roots cool, and prevents soil compaction. Finally, it serves as a home and food source to millions of earthworms, beetles, and grubs, as well as trillions of microscopic soil organisms, all of which are instrumental in helping litter (twigs and leaves) decompose into nutrients for plants.

Covering the soil with an organic mulch is the first step to creating a low-input, self-supporting system. Exposed soil is a welcome bed for weed seeds that germinate easily and grow in such conditions. Compaction is another problem. When objects as large as gardeners or as small as falling raindrops tread on bare soil, the surface gradually becomes compacted. This can transform fluffy, porous soil into a surface like concrete that prevents water and air—two elements that roots need for plant survival—from penetrating. Lastly, the absence of an organic soil cover means there is no raw material that can be broken down into plant nutrients.

By mulching the garden with woodchips, dried leaves, grass

and other plant materials, the gardener creates a spongy layer like that found in the grasslands and forests. This layer is broken down slowly by soil organisms and builds soil fertility. Mulch prevents the soil from drying out in summer. It keeps plant roots warm in the winter, and helps suppress competing weeds. Simple woodchips are one of the most important materials used for keeping the garden productive and the environment healthy.

Donald and Daisy, our Peking ducks, take a leisurely swim in the gazebo pond.

In creating an environment that helps control or eliminate insect pests naturally, one avoids toxic and harmful pesticides that can pollute the environment and also kill beneficial insects necessary for any healthy garden. Birds and insects such as ladybugs, wasps, the praying mantis, lacewings, and the big-eyed bug eat many insects that are harmful to garden plants without upsetting the garden ecology.

Other insects, such as butterflies, not only add beauty to the garden, but, along with bees, are important pollinators that fertilize fruits and flowers for seed production. By understanding the life cycles and food sources of beneficial insects, one can plant the types of flowers, shrubs and trees that support their population and create a more dynamic and sustainable system.

Biodiversity in the Garden

The more variety of plants in a garden, the more natural checks there are on insects and plant diseases. For example, plants such as marigold and lavender emit chemicals as a defense against pests, and can help prevent the movement of insects, fungus or diseases from plant to plant that would soon infest the entire garden. Birds need thick shrubs or trees to build their nests; planting trees near or in the garden for birds, which feed on insects, can be a major pest control. Large and

An early summer view of the bird and butterfly garden with flowering sedum, yellow hardhats (*Centaurea macrocephala*), catmint (Nepeta grandiflora) and goat's beard (Aruncus dioicus). The tall tree in the background is a serviceberry (Amelanchier canadensis).

small perennial flowers, grass, diverse trees and shrubs provide food and shelter for wildlife through the winter, increasing the biodiversity of the garden system.

Humans have long played a decisive role in shaping their environment to promote diversity. In New England, the Native Americans controlled the thick stands of old growth forest with periodic burns. This created a mosaic of meadows and forests in different stages of regrowth, which in turn supported greater numbers of game for them to hunt. Many of the "pristine" forests in the Central American jungles are nothing more than overgrown kitchen gardens established by the Mayan Indians centuries ago.

Our ancestors realized, even then, that diversity in gardening was a prosperous route to survival. It has only been with the rise of industrialized technology that humans have altered their environment in ways that now threaten the long-term stability of the earth's vital systems; its water, air and soil. Promoting healthy land use in sites such as prison gardens helps inmates understand the greater need to be effective stewards of our planet's resources; it connects them to their role as caretakers of a fragile and often threatened system.

For students in the program, "natural systems" gardening is a fast track introduction to organic farming, the need for healthy, fresh food and the importance of keeping their own

The Story of Alcatraz: A Return to Nature

If working with landscape is thought to socialize inmates, could the same be said of the inmates' impact on landscape? One needs to look no further than Alcatraz, the former maximum-security prison built on an eleven-acre island in San Francisco Bay, for an appropriate answer. Here, amidst buffeting winds, cold fog and the occasional storm, three generations of wardens, together with officers and inmates, undertook the task of transforming a bleak, rocky outpost into a contoured landscape of gardens and flowers. Dynamiting steep rocky slopes and carting in tons of soil from the California mainland, the residents carved out terraces to cultivate a site that once was nothing more than a bird and seal rookery.

left: The formal Victorian rose garden at Alcatraz in 1870, when the island was a military prison. right: Today, formality has succumbed to the wild impulse of nature as seen by the red valerian that has taken over the long abandoned officers' quarters.

Established as a military garrison in 1853, Alcatraz gradually grew into a military prison, becoming a federal penitentiary in 1933. The creation of gardens made the island somewhat bearable for staff and their families. Victorian gardens were initiated during the Civil War, when the island was peopled with rebel sympathizers and soldiers, Union deserters and captured Native Americans. In 1915 the army took a big step in prison reform by developing a progressive vocational program aimed at prisoner rehabilitation. In 1917 the Alcatraz newsletter, "The Rock," reported that "a training program had begun that provided eight men with experience and skills to become gardeners."

Across the island, inmates cut deep holes in the rock, filled them with soil and planted roses, sweet peas and lilacs. For the next 50 years, long after the island became a federal penitentiary, both staff and inmates played a dramatic role in expanding the plantings and gardens on Alcatraz.

One of the more influential gardeners on Alcatraz was Freddie Reichel, secretary to Warden Johnston from 1934 to 1941. Reichel was taken by the army's efforts to beautify the island and soon began to spend his free time maintaining the existing beds and expanding the gardens. He took over the "rose garden, the greenhouse, the slope behind his quarters, and the small flat garden near the post office. Reichel brought with him to Alcatraz little horticultural experience, but by the time he left he was an expert self-taught gardener. Isolated, yet no less than a half-mile from the San Francisco, he enlisted support from some of California's top horticulturists."

The prison, closed in 1963, soon deteriorated into a complex of gutted concrete buildings, broken asphalt and half-standing pitted walls. But what nature destroyed it also reclaimed. The carefully tended gardens overran their terraced walls, their seeds carried by wind and their berries spread by birds. Habitats emerged. Nature quickly reclaimed the island for itself. Alcatraz had been transformed once again, and is still transforming into a dynamic community of animals and plants, an evolving system of native and naturalized exotic trees, shrubs, perennials, annuals, grasses and ground-covers. ■

bodies free of harmful chemicals. It is obvious to anyone who has spent time in jail that years of low nutrient, fatty diets, along with tobacco and hard drugs, have had a tremendous impact on the health of inmates. At the same time, the Center for Disease Control estimates that diseases such as AIDS, Hepatitis C and tuberculosis are as much as seven to ten times higher than the general population. Poor ventilation, close quarters, at-risk behavior, starchy mess hall foods, and sugary commissary snacks all exacerbate the pre-existing health conditions inmates enter prison with.

Riker's students learn how to organically grow vegetables, which are then consumed at the greenhouse or distributed to local soup kitchens.

I often tell students to think of the garden as a reflection of their mind and body. Chemical fertilizers, herbicides and pesticides may generate high yields, but create a dependence that in the long run will damage the garden along with the surrounding landscape. Students need to cultivate positive relationships in their family and community, in the same way organisms benefit each other in a healthy functioning system. Unhealthy relationships or negative behavior should be removed in the way weeds and pests are removed without using harmful methods. Healthy inputs such as compost, which takes organic debris and waste and converts it into fertilizer, is simply a metaphor for dealing with mistakes in one's life and using them as a basis of experience and wisdom. I tell them that personal development, like gardening, is a process that takes work and discipline, but over time generates its own rewards as freely as the fruit and flowers of a well-tended, self-sustaining garden.

CHAPTER SEVEN

Building Curriculum

Over 1.5 trillion dollars was spent to fight the drug war between 1970 and 2000.

—GEORGE WINSLOW, BUSINESS JOURNALIST

If we wanted to stop drug use in the United States there's an easy way to do it: educational programs.

—NOAM CHOMSKY, PROFESSOR AND AUTHOR

The Rikers GreenHouse curriculum develops skills in horticulture while building a foundation in plant science and ecology. Because students revolve through the GreenHouse throughout the year, the curriculum must be flexible and take into account the seasons, the activities taking place in the garden, and the skill levels of the students. Certificates of achievement are awarded based on the number of hours in class and in the garden, and on the completion of a project specifically relating to the student's interest in horticulture or construction.

There are a number of horticultural programs, specifically in state penitentiaries, that are conducted through state extension agencies or community colleges in which students receive certification through the Nurserymen's Association. These technical programs work best with long-term inmates who have at least one year in the same facility. They cover such topics as turf management, pesticide use, propagation, pruning and greenhouse management.

At Rikers students are required to have a basic classroom understanding of plant and soil science and their practical applications in gardening and landscape management. In two months of instruction, students are expected to answer the following questions that are key to a working knowledge of horticulture:

Botany

- What is a plant?
- How do the leaf (photosynthesis), roots (nutrient and water cycles), stem, (phloem and xylem cells, plant growth), and flower (sexual reproduction, fruiting, seed dispersal) support the plant physiology?

Soil Science

- What is soil (organic matter and minerals)?
- What is soil texture (clay, silt and sand)?
- What are soil horizons (organic matter, topsoil, subsoil, parent material)?
- What is good soil for plant growth (sandy loam, well drained)?
- How do you determine your soil texture?
- What animals are important for healthy soil?

Soil Amendment and Care

- How can you amend your garden soil?
- What is decomposition and compost?
- What is pH and how can you change the level of pH in your landscape?
- What is mulch and why is it important?

Elements of the Garden

- What is a tree (canopy and understory, deciduous and evergreen/conifer)?

- What is a shrub (broadleaf, evergreen, deciduous, spring or summer flowering)?
- What is a perennial, annual, bulb, vine, lawn, ornamental grass and groundcover and how is each used in the landscape?

Ecology

- What is the relationship between natural systems and garden systems?
- What are plant/animal communities and interactions?
- How do we replicate nature in the garden?

Experiential Education

I have found that students become more engaged and learn better through hands-on instruction rather than structured classroom learning. Although students are required to attend lectures and take exams testing their knowledge in each subject, I try to limit class time and focus on activities in the greenhouse or gardens. Many of the lessons taught in the classroom are reintroduced and reinforced as students go about their work pruning trees, starting seeds, watering, controlling weeds and pests, spreading compost and all the other tasks related to gardening. It may sound like a cliché, but instructors need to make learning fun if they want to be effective in their classroom.

Low literacy is the norm in jail. Of the 350 inmates who have passed through the GreenHouse program, less than 30% had a GED or high school diploma, only 10% ever attended some college, and less than 3% had a college degree. Education is a priority, but most inmates are the deficient product of a traditional school system. A high percentage, roughly 30% in our program, had a learning disability and required some form of special education as children. Our data is commensurate with national findings where 19% of adult prisoners were found to be completely illiterate and 40% were functionally illiterate compared to 4% and 21% of the general public.

A view of the vegetable garden in winter. Inside the fence are plants donated by nurseries that will winter out at Rikers before finding their way to GreenTeam projects in the city.

In the winter months, when work is mainly confined to plant care in the greenhouse or to carpentry and construction projects, we conduct workshops in math, writing, computer literacy, drawing and landscape design. The workshops are designed to help students focus on subjects that are applicable to everyday life or specific projects at the greenhouse, and are often taught by guest speakers who are noted professionals in the field of ethnobotany, ornithology, landscape design, wildlife rehabilitation, ecology, horticulture and journalism.

The importance of developing good grammar and writing skills cannot be understated. Most inmates communicate with

Garden Math: Formulas for Estimating Numbers of Plants

Michael Ruggiero of the New York Botanical Garden developed a system that approximates the number of plants that fit into a planting bed. Instructors can use this as a hands-on lesson in geometry, as well.

Step 1. Determine the number of square feet in the area to be planted.

- For rectangles multiply length by width
- For circles multiply the radius of the circle by itself then by pi (3.14)
- For ovals multiply the average radius by itself then by pi
- For triangles multiply 1/2 the height by the base

Step 2. Determine the number of square inches in the area to be planted by multiplying the number of square feet by 144 (which is the number of inches in 1 square foot).

Step 3. Determine the number of square inches a mature plant will cover. (This can be accomplished by multiplying the suggested spacing between plants by itself).

Step 4. Divide the number of square inches required for one plant into the number of square inches in the plot. The answer will be the total number of plants needed for that plot.

lawyers, judges, and caseworkers through letters, and their ability to effectively construct their thoughts in well-written language can have a significant impact on their case. These skills serve them well after their release when they search for a job or enroll in school.

Each class is responsible for the production of a newsletter or journal that details students' experience in jail or at the greenhouse. In the past, students have written articles about plants, design, nutrition or finding jobs after release. Describing their experience in an article helps students with their writing, grammar, computer literacy and art skills. The newsletter not only provides students with a showcase for their work, but helps document the program, informs correction staff and officials about the program, and provides promotional material for public events and fundraising.

Rikers Island: The Gazebo Garden

NOW IN SPRING, the gazebo and pond stand by themselves in a flat grassy field. The past winter, inmates had designed the area placing the gazebo in the center of a garden. The pond would be partly concealed by tall ornamental grasses. A walkway through the garden and alongside the pond would connect the gazebo to the classroom, moving the visitor through trees and a perennial garden to the small staircase ascending the structure. With the snow melted and temperatures climbing to tolerable conditions, four men take on the physical task of removing the grass and building the garden.

Working from a master plan they marked out the area with wooden stakes and string. Using mattocks and shovels they turn over and rake out mounds of weedy lawn. It is backbreaking work, and on more than one occasion a student mentions that in the 20th century, roto-tillers make slave labor easier. I counter that doing it by hand will make them appreciate the garden more once it is completed. That answer is usually greeted by grunts, so I tell them not to think of it as labor as much as a learning experience, like school. When they tell me they dropped out of school precisely for that reason, I change tack and go for the machismo angle, saying I'm doing the same work they are and prefer the physical labor to a desk job in an office. When I'm told they would also enjoy the work like me if they were paid the same amount I was, I resort to the standby argument they never have an answer for: if you don't like this kind of work, don't come to jail. Silence.

It's good to dig by hand. I have experience on my side. The gardens on our two-acre site were carved from the compact, gravelly earth in this fashion, and each inmate who contributed expressed the same feeling of pride looking at the finished product. At those moments it is hard to resist mentioning the clichés of hard work and good results, or nothing of greatness ever comes easy. It is precisely those moments that produce life-altering insights—the realization that by making a plan and sticking to it regardless of how backbreaking, long and tedious the work, they can achieve their goals of building a successful life. Hard work, I tell them, is part of the process. Be proud of your labor.

As the men head inside for their ritual coffee and box of Frosted Flakes with milk, I pull Jason B. aside and tell him that he needs to mark the gazebo pathway. Let it curve into the gazebo. Flare the entrance so you feel you're entering the garden then narrow it so it eventually is the same width of the gazebo staircase where it ends. Jason understands exactly how I want it. On the site design the students have drawn a pathway of equal width. But as the garden proceeds, the design alters to accommodate artistic flourishes. Jason agrees. "A pathway with a wider entrance seems more inviting," he says. Jason is barely twenty years old, but in this particular group I rely on him to translate ideas into actual work in the garden. More than anyone, he has the wonderful capacity to take plans and improve on them. Perhaps it is intuitive: Jason, an Hispanic kid from the Bronx with no prior gardening experience, has an eye which discerns shape and texture in the landscape. Despite his limited knowledge he knows what works, what plant goes where, which tree should accent a space in the garden, the direction and flow of a well-built path and the most effective color of foliage or flower to offset the next major element in the garden.

If that weren't enough, this tall, good-looking boy has the innocent charm of a freshman in a college fraternity house. He can charm the officers—both women and men, the civilian staff in the school and library, his legal aid lawyer, guest instructors and me. Perhaps his former vocation had an influential role on his way around people. When he wasn't selling drugs, Jason was a professional clown, making $75 an hour hosting children's parties in the Bronx and Brooklyn.

"To be successful as a clown," Jason tells me, "you have to get those kids to like you. It doesn't matter how funny you are, how good your tricks are, what you do. If they don't like you, they'll tear you apart."

"How do you do it?" I ask him. "How do you get on their good side?"

"First, you gotta be real," he answers. "Then you have to give them what they expect. No surprises. Little kids in the *hood* don't like surprises," he explains. "They got enough in their life to deal with. If they expect animal balloons, make sure you give them animal balloons. The girls want the *macarena* contest. That's what they did at the last party and that's what they expect from this one. Keep them moving. If they sit too long and get bored you're a dead clown."

With skills like these, Jason can survive in jail. More importantly, with these skills he can survive outside jail. Despite being abandoned as a child and spending time in foster care, Jason managed to get through school earning a GED before he was eighteen. His girlfriend, and mother of the child he fathered at the age of sixteen, worked in a bank. Before his arrest they shared a basement apartment in his old neighborhood in the Bronx, a relatively stable life for someone who grew up in the unstable and often terrifying environment of the foster care system.

"You don't need to be dealing drugs for a living," I tell him.

He agrees. "I want to be the first in my family to go to college," he says. "I want to be an accountant. I want Jason Jr. to be proud of me."

We measure off the front entrance of the path. With two

stakes, a mallet and a ball of twine, we mark off what will be the borders to the path. We are joined by Hector B. and Tyrone M. Hector takes the stakes and every five feet, where Jason tells him, plants one in the ground. Hector is a hard and uncomplaining worker who learned landscaping at a penitentiary upstate where he spent three years serving time for robbing a *bodega*—a Latino neighborhood convenience store. Now he was at Rikers for a drug sale. "A bogus charge," he told me. "They pulled me on a sweep, then got me on an old charge." He shook his head. "Once you're part of the system it's hard to get out."

He bangs the stake with the mallet while Tyrone begins tying string from stake to stake. Despite his interest in gardening, Hector, unlike Jason, has a strong resistance to learning anything new or academic. "Just tell me what to do, and I'll do it," he told me. At first, I'd ask him where to plant a specific shrub or tree, or where to mark the garden border and he'd just look at me and ask where he should dig. It was as if he deliberately resisted taking the responsibility of making a decision. After a while I gave up. Hector was stubborn. If we sat in class for a lecture he would stare slackjawed, completely absent from any discussion. Often he would fidget, then walk outside for a trip to the bathroom. Or stand at the back of class and feign interest. Then pace. I once asked whether he learned anything from lectures. "I can't sit for long," he told me. "Just give me something to do."

Hector turns the lawn and rakes it into piles that are carted off and buried for compost. The dynamics between inmates develops rapidly. Jason, almost fifteen years younger than Hector, knows he can not assume a supervisory role or give him specific tasks. But he quickly learns how to manipulate him to do most of the labor. It can take many forms, but Jason's most common tactic is to initiate a physical job, such as turning over the lawn, moving compost, or hauling rocks and then feign an apparent struggle. Hector immediately assumes control telling Jason he's not doing it right. Jason then moves to a more cerebral, less strenuous project. Not that Hector is fooled. He knows exactly what Jason is doing. It is more of a silent and mutual agreement: *Don't be a punk and give me orders, and I'll do the work.*

When the string is tied and the sides marked off, I explain the next step is to remove most of the soil inside the path to a depth of six inches. Once that is complete, the path is raked smooth and the remaining soil dumped into the surrounding beds. Hector, Tyrone and the two new men are to bring in sand. The sand is to be spread on the path to replace the soil, and again raked smooth at a depth of four to five inches. The last step is to fill the path with stone.

Almost one month previously, Officer Periera had discovered a large cache of smooth brown and white construction pebbles on the far side of the island along with piles of sand. The men, prior to arriving at the greenhouse, would stop at the site and fill up plastic garbage cans with sand and stones and bring them to the garden. The path to the gazebo would be made of these pebbles, an attractive material that would complement the wood, stone and water that comprised the waterfall complex. This was typical of most of the design and construction of the greenhouse gardens. What we designed, built or planted was entirely dependent on the materials we salvaged or had donated to the program. If Periera hadn't found the pebbles we would have used crushed brick, gravel or worse, woodchips, from which sprout a profusion of weeds, making a path more like a hiking trail. In this case, we got lucky, and, as I tell the students, "it's making do with what we find, but we won't find it if we don't look. It's a bit like life," I can't resist adding.

We break for lunch. Inside Hariberto M. has dished out a concoction of rice and beans slathered with potatoes covered in garlic and cilantro. "It's not much," Hariberto says, "but it beats mess food." All the men agree. Food is a big benefit at the greenhouse. Unlike most upstate penitentiaries, the men do not have cooking facilities inside the jails at Rikers. They either eat whatever's served them at mess, or eat the pre-packaged food they buy at commissary. I supplement their diet with staples of rice, beans, onions, pasta, and chicken and whatever canned or frozen foods Periera is able to scrounge from the jail. In the summer, we harvest our own vegetables and herbs. We eat salads and make our own dressing and teas. This brings the quality of jailhouse cuisine to a new level. It also gets our students thinking—with a bit of prodding—about fresh food, nutrition and diet and its impact on their physical health. An informal and rather unscientific survey I've conducted over the years revealed that only 30% of my students has ever eaten an eggplant, let alone seen it grow on a plant.

Building the gardens is a continuous process; new space carved out from the weedy lawn is ready for planting.

I once visited a garden project run through a federal penitentiary in Staten Island. There, inside the recreation yard adjacent to the housing units, the men had a small greenhouse and surrounding garden where they grew herbs and vegetables. Every Wednesday they had a mini-farmer's market for the other inmates who would line up to buy fresh greens, jalapeno peppers, eggplants, potatoes—whatever was harvested that morning. Proceeds from the market would cycle back into the garden for seeds and tools. Some of the men had been running the garden for over a decade, enough time and experimentation to develop into very good gardeners.

After lunch I ask Jason whether we need a barrier along the path to keep people from cutting through the garden on their way to the gazebo. I tell him I like the way the stakes and string look forming the path; perhaps we could find something more finished and permanent. While Hector and Tyrone scoop out the pathway, I tell him to design a guardrail.

"You want a stronger statement here," I tell him, "something that delineates the path from the garden, and something that prevents people from trampling the garden to look in the pond." Inside the shed where we store our tools, supplies and carpentry equipment, Jason notices a quantity of thick nautical rope and suggests he base his design around that.

"Rope is the way to go," he says. "In jail, people use it to either escape or kill themselves. We put it in the garden to mark pathways. What better statement is that? "

Learning by Building

Vegetable gardens, orchards, perennial walkways, herb and rose gardens, ponds, a restored native system, memorial gardens and massive plantings are a striking array of projects that can help inmates strengthen their knowledge. Each project is an opportunity to add another block to the program's curriculum, enhancing the level of skills students leave jail with.

The Vegetable Garden

No garden is complete without space for growing food. Vegetable gardens are an important introduction to horticulture; not only are the rewards of planting and maintenance realized in a short time, but fresh vegetables are a healthy addition to inmates' diets in jail. There are also intangible benefits.

When students in the program were asked how they felt about eating produce from the fifty-by-thirty-foot garden they tended at the greenhouse, they mentioned things like self-sufficiency, knowledge, a connection to the earth, pride, satisfaction, and a feeling of importance.

It is not a stretch to state that vegetable gardening has all the beneficial elements of a therapeutic program. Life metaphors abound in the rituals of preparing soil, sowing seeds, the transplanting of seedlings, thinning and weeding, fertilizing and watering, and of course, the eventual harvest.

Certain programs stress the advantage of communal work in facilities where inmates have short sentences and may not spend an entire plant-to-harvest season in the prison garden.

Curriculum for a Kitchen Garden

A science and horticulture curriculum can be based entirely on a small annual vegetable garden. The following is one example of an integrated science education, life skill and vocational work curriculum:

Soil Preparation	Qualities of good soil and how to improve it. The importance of pH and how to test the soil.
Composting	Decomposition and nutrient recycling.
Seed germination	Reproduction and plant growth.
Plant growth	Photosynthesis and transpiration.
Thinning	Plant competition and ecology.
Weeding	Invasive plants and their control.
Watering	Evapo-transpiration and the water cycle.
Pests	Natural "integrated pest management (IPM)." Predator-prey relationships in the garden.
Harvest	Plant reproduction and fruit production.
Recipes	Nutrition and its effect on human health.

Six-Week Woman's Program at Suffolk County Prison

Cornell Extension has established a six-week course and "kitchen garden" for women inmates at the Suffolk County Correctional Facility in Riverhead, New York. According to program director and instructor Darlene Widirstky, it is the practical application of what the students learn in the classroom that has turned inmates into gardeners. Produce from the garden is donated to Long Island Cares for distribution to soup kitchens and food banks throughout Suffolk County. For Widirstky, the highlight of her program came when a women in maximum security claimed, after completing the six-week program, "this is the first thing I have done successfully."

Each subject is provided on a weekly basis:

- **Soil Science:** A study of different soil types and the importance of soil testing. Changing soil to suit plants' needs is covered under soil amendments.
- **Botany:** An overview of plant biology and taxonomy. In addition, types of plant life are covered, i.e., annuals, biennials, perennials, vegetables and herbs; as well as propagation and seed starting.
- **Integrated Pest Management—IPM:** Different environmentally-friendly ways to manage pests, including companion planting.
- **Trees & Shrubs:** An overview covering proper planting methods and the maintenance and care of woody plant material.
- **Garden Types & Design Elements:** An overview of selected garden types and the importance of a well thought-out design.
- **Cultivation, Maintenance & Plant Record Keeping:** Students establish and maintain garden areas so they apply what they have learned in a practical way.

—The Cornell Extension Program, Suffolk County Farm

The idea of a common bed reinforces the spirit of shared responsibilities and common goals in the work place. The motivation is simply being part of a process, as opposed to actually reaping the rewards of a productive harvest.

On the other hand, providing inmates with individual plots enables them to make decisions that reflect personal or cultural expression, creativity and shared responsibility for maintaining a productive garden. Inmates with long-term sentences may be better off with individual plots. The continuity of gardening allows the student to learn through years of trial and error. Finding out what works is a process of incalculable value.

Vegetable gardens can be culturally thematic with sections cordoned off for plant foods from different regions around the world. Caribbean plant foods can reside near a plot for African foods, or herbs from Southern Europe. There are limitless ideas to bring students closer to geography, the diversity of plant communities and human culture. At Rikers the instructors devoted a small space in the garden to growing the "three sisters," a Native American companion planting scheme involving squash, beans and corn.

The positioning of beds can be as creative or pragmatic as space allows. Generally, raised beds seem to work best

"Three Sisters" Curriculum

The Three Sisters is a Native American planting scheme that integrates beans, squash and corn in a mutually dependent gardening system. It works like this:

The bean plants wrap their way around the corn using the sturdy corn stalks as support. The bean plant provides the corn with nitrogen for the next growing season when the spent leaves are mulched into the soil. The squash vine wraps around the base of the corn creating a dense mat that represses competing weeds. The vine also creates a humid micro-climate which helps retain moisture at the soil surface while the tough spines help repel hungry predators like raccoons.

The importance of this elegant system was not lost on the Iroquois who made Three Sisters a part of their tribe's creation myth, and for whom the crops were their nutritious and dietary staple.

Instructors can use this system as a lesson plan that introduces concepts such as nitrogen recycling, mutually beneficial plant interactions, micro-climate, plant architecture and ethnobotany. Three sisters also demonstrates ways to maximize crop production in a low maintenance, energy efficient system that is sensitive to the surrounding environment.

It should be noted that certain varieties of corn, squash and beans are more suited for this system than others. Runner beans, either vine or pole varieties are needed to climb onto the corn, while the corn needs to be a sturdy type that can bear the weight of heavy-laden bean vines. The squash must be a variety that forms a dense ground cover to serve its purpose as a living mulch. ■

because they demonstrate the demarcation between well-tended soil and public garden footpaths. Sandy soil, which generally is not suitable for raised beds, should be well amended with compost and manure. Over time, the beds will benefit from a continuous source of organic amendments, not only increasing the available nutrients in the soil but also improving the soil's texture and its capacity to retain moisture.

We encourage diversity and randomness in the vegetable garden in the same way we look towards diversity in the "natural garden." For example, a small plot consisting of tomatoes inter-cropped with marigolds, lettuce, beans and basil ensures the beginnings of a colorful, fragrant, early-to-late summer harvest with a built-in pest control system. Students should be encouraged to develop their own combinations for systems that are productive, have long-lasting harvests and are aesthetically suited for public spaces.

Finally, planting a mix of annuals and perennials that have edible flowers is a great way to enliven food with taste, nutrition and an aesthetic garnish. Nasturtium, Johnny jump-up, violet, and pansy are flowers that provide nutrition and color to a summer salad.

The Herb Garden

We have found that our herb bed, more than any other garden on Rikers Island, stimulates the students' interest in plants and gardening. Imagine wandering past aromatic plants that can help cure headaches, heal wounds, sterilize cuts, clear up acne; that can be made into fragrant soap or candles, relaxing teas and culinary spices; that can relieve gas, help with indigestion and boost the immune system.

"It's an outdoor pharmacy," a student once remarked during a class on herbal remedies. "And it's free." For inmates who have difficulty acquiring simple drugs such as aspirin, and

The herb garden in late spring is a source of food, spice, tea and raw ingredients for greenhouse products such as natural-made lotion.

Learn About Herbs: Making Fragrant Soaps and Lotions

During the cold months of winter we conduct workshops developing different products from herbs including lotions, lip balms and soap. Using beeswax as our base, lotions are easily created from fragrant oils of lavender, citrus, rosemary, bergamot, and mint. An infusion of dried flower petals from calendula, vitamin E and bee pollen is added for its antiseptic and restorative qualities. The process familiarizes students with the kinds of beneficial herbs grown in our gardens, their historical and curative uses, along with the benefits of products derived from bees. We even created our own label as a study in product marketing. Detailed recipes using these ingredients are readily available on the internet and in herbal books and magazines.

NOTE: Fragrant lotions are a hot commodity in jail where skin cosmetics are limited or viewed by Department of Correction officials as contraband items. At Rikers, the officers allow students to use what they produce in the greenhouse only. Students caught bringing lotion, balm or herbs back to their housing unit are promptly fired from the program. ■

The Taste and Smell Test

A way for students to identify many herbs is to have them crush and smell the leaf and familiarize themselves with different odors. Is the smell sweet or pungent? Does it smell like candy or citrus, or remind them of cooking food? With a sweet smell, how does it taste? Smelling and tasting is one of the first steps in developing an ongoing dialogue with familiar and unfamiliar plants. Perhaps it invokes a primordial urge to use the senses in exploring the unknown and unfamiliar world as a potential food source. As one inmate said, "If it smells, it must be used for something." ■

whose toiletries are generally of poor quality, herbs are especially appealing.

With minimal space, the prison herb garden can produce enough mint, basil, lemon balm, fennel, oregano and thyme to fuel a small cottage industry, or at least supply the facility with culinary spices. Herbs are generally low maintenance crops that are drought hardy and often perennial. Mint, for example, will spread under most conditions and can be used in tea, salad dressing, garnishes and cooking. Lemon balm, an invasive perennial, makes a soothing tea for headaches and a salve for bug bites and infections. Basil, though an annual, responds to frequent harvests through the summer, and its fragrance and amazing number of varieties make it an interesting crop.

Because of their ability to attract beneficial predators or their capacity to emit chemicals that ward off insect pests, herbs are important companion plants for vegetables.

At Rikers, our students designed, measured, and installed an herb garden employing skills that combined their knowledge of plants with design principals and personal aesthetics. Culinary and healing herbs were chosen for their productive qualities. Lavender was inter-planted with thyme, parsley, mint, hyssop, strawberries, rosemary and garlic. The circle or center of the garden was planted for color and food with salvia, marigolds and chili peppers. This garden bears more than just fruit, or, in the case of herbs—leaves and stems. The garden gives students an understanding of human history and the commercial uses of botanicals.

Gathering and Storing Herbs

Gather herbs on a sunny day when completely dry. Flowers should be fully open.

For Flowers	Pinch off or cut off flowerheads. Place the flowers in a paper bag, tie loosely and hang in an enclosed place.
For Leaves	Pick leaves and stalks, tie them together in a bundle and hang them upside down in a dry airy place out of the direct light. Once the leaves are dry, strip them from the stalks and crumble them up. Store them in a dark airtight container such as a brown bottle.
For Seeds	Cut the flower stalks in fall when the flowers are well desiccated. Bundle, tie and hang stalks upside down. Catch the seeds in a paper bag tied around the stalks.
For Roots	Dig up roots, separate from the plant and soak in cold water for about one hour. Cut into small pieces and place in a paper bag in a warm dry place until thoroughly dry. Store in an airtight container.

The Butterfly Garden

If gardening creates an on-going drama, then butterflies are the principal actors accenting their role with movement and color. They are not only beautiful, but contribute to the fertility of the garden, being second only to bees as pollinators. Unlike formal gardens with their fragrant roses and flowering hybrids, butterfly gardens are reliant on a jumble of flowering plants that provide nectar for the butterflies all season long, as well as habitat and food for their larvae, caterpillars. A small flowering plant hovering in the corner of the garden may be considered a weed, but it also might be an important food source for a fritillary, swallowtail or monarch. Attracting butterflies to the garden, especially in isolated areas like the yard of a prison, demands a greater understanding of what butterflies need to survive and proliferate. What do butterflies eat? Which kind of plants do they lay their eggs on? What does the larva need for survival in each stage of metamorphosis as it transforms into a chrysalis and then a butterfly?

Butterflies also need windscreen plants to shelter from the wind, sunning spots (stones for example), and water for drinking. Building a small rock pool is a worthwhile project for students. The pool should be lined with plastic to prevent the water from draining but be shallow enough to evaporate the water before it stagnates with mosquitoes and algae.

It is important for students to identify caterpillars to determine whether they are a garden pest or a desirable butterfly-in-waiting. Spraying a bacterial pesticide on what might seem to be an area abundant with striped parsley worms munching on dill and carrots would wipe out the beautiful black swallowtail. Once properly identified, caterpillars can be removed by hand and placed in another part of the garden—a spot that could be planted specifically with their food and used as an exclusive feeding ground.

Attracting Birds

Birds are not limited by the high walls and barbed wire of prison and are therefore attracted to a well-thought-out "natural garden." Their sounds and color are a source of life that adds dimension and value to the inmates' gardening experience. With the right selection of plants for food and cover, prison gardens can provide valuable bird habitat, something that is quickly disappearing in the U.S.—meadows, marshlands and young forests have given way to suburban development. Even in heavily landscaped areas and homes, plant

A Note about Planting: Death to Air Pockets

Experience has demonstrated that the greatest risk of mortality for newly planted perennials and shrubs is poor planting. On many occasions while examining a dead or wilted plant I have watched my hand disappear easily between the root ball and newly dug hole.

Rule #1

Make sure the soil is firmly packed between the root zone and the hole. Air pockets prevent the roots from making contact with the soil peds, which draw water through the soil in a capillary fashion. Without that contact the roots virtually dry up, killing the plant if there are too many pockets in the soil. Students should use their hands to firmly press the soil into the hole, as opposed to just moving soil into the hole with a trowel or shovel.

For heavy clay soils, avoid walking around the newly planted hole to avoid compacting the soil and preventing water and oxygen from penetrating the surface layers.

Rule #2

Always remember to water after planting, giving the soil—not the plant—a good drenching for the roots. ■

choices tend to favor hybridized shrubs and flowers that provide large blooms at the expense of edible fruit for birds (hybridized plants tend to be sterile), not to mention the replacement of meadow grasses with sterile lawns.

Growing plants and trees that provide food and cover is a requisite strategy for attracting birds to the prison garden. While the landscapes inside most prison grounds tend to be spare, small evergreen shrubs and hedges can be a major source of cover for nesting birds. Native species should be used at any given chance including varieties of *Ilex*, *Viburnum*, *Amelanchier*, *Juniperus*, and *Prunus*. Climbing vines such as species of grape and Virginia creeper are important food for birds, along with the wealth of protein found in the myriad insects attracted to a highly diverse multi-structured garden.

To attract year-round residents, feeder boxes are a necessary addition to the prison landscape. Students are encouraged to document the different species of birds visiting the Rikers greenhouse. This activity mirrors *Feeder Watch*, a program established and run by the National Audubon Society, which encourages homeowners, schools and different institutions to track and monitor bird populations from feeders in their backyards. Instructors can involve their prison gardens with *Feeder Watch* and contribute data to the nation-wide tracking program by simply writing to the Audubon Society and requesting a *Feeder Watch* packet.

The Arbor

The silhouette of deciduous trees in winter or the deep green of conifers in the snow, the flowers of spring, the fruit of summer and the foliage of autumn make powerful statements in the prison landscape. Trees not only lend a dynamic structure to the garden, they are ecologically vital for maintaining a diverse plant and animal community.

The pagoda in summer covered by grape vines creates a private space for a meditative moment.

Trees also provide a symbolic link between inmates and the slow evolving nature of time. It is this connection that allows inmates to interact and reflect on life and its consequences in a powerfully new way. Planting a sapling that will reach maturity in 80 years time, or a tree that will one day stand 90 feet or more is an overwhelming concept to students who often view life in snapshot moments. Trees imply commitment and slow growth. The act of planting a tree is a responsibility that will have effects on the environment for the next century, and that impact is transferred to the inmate's own experience of change, growth and maturity.

But trees take planning. Large growing or full-grown trees

Fall in the pagoda garden illustrates the beauty in the change of seasons and the distinct pattern of change and renewal.

How Does a Tree Grow?

In a lesson about tree growth, we pose a riddle for students to ponder: A twenty-foot tree has a painted stripe circling the trunk at five-feet above ground. Thirty years later the tree is sixty-feet tall-where is the stripe? Trees grow at apical meristems from the tips of branches, elongating them. From these new shoots come new leaves which help the tree manufacture food from the sun. This is called primary growth. Lateral meristems, found along the length of the stem, add girth-sometimes an inch of thickness each year for mature trees, in a process called secondary growth. As the tree grows, the lower branches receive less sunlight and die, eventually falling off. This gives the illusion that the tree is thrusting the lowest branches ever higher as it moves upward.

A majority of students work out complex math to determine the height of the stripe, but the simple answer is, of course, that the stripe remains at five feet. ■

can impede or block site lines, which in prison is often viewed as a security risk. Under wires, eaves, next to fence lines, near buildings where windows could one day be obscured are sites that should be avoided.

Avoid Large Trees

A possible way to integrate trees in the prison garden is to set aside an area for a small arbor using understory species that do not reach the dramatic heights of large canopy trees. In temperate regions, such as Rikers Island, the arbor can consist of dogwoods (*Cornus spp.*), redbuds (*Cersis*), serviceberry (*Amelanchier*), magnolia (*Magnolia spp.)*, mountain ash (*Sorbus spp.)*, dwarf junipers and cedar (*Juniperus and Chamaecyparis*), to name a few. It is also our experience that planting fruit-bearing trees (dwarf varieties) is effective in stimulating the students' overall interest in trees and tree care. The reward of course is seasonal, appearing in the edible form of delicious fruit. A mixed planting of pears, peach, apple, apricots, plum and cherry can yield a stream of fruit all summer long and create year-long interest with winter maintenance, spring flowers and fall color.

The mosaic pathway leads past the edge of the nature forest and fruit orchard.

Overall, the arbor provides students with a prevailing opportunity to learn about and develop skills in tree care. For inmates with an interest in horticulture, the importance of these skills cannot be understated. Tree care is a year-round multi-million dollar business in the U.S. and offers a variety of well paying jobs for climbers, pruners, surgeons and planters. Maintaining a small arbor integrates a number of skills that inmates can later use for employment or for certification through professional organizations.

Turf management

At Rikers we do very little with lawn management and care. Most of our experience is filling in ruts or broken turf with a bit of topsoil and seed and mowing. By summer most of the island's green turf is brown, dependent on a summer rain rather than a well-timed sprinkler. But there is much value in learning about lawn care. After all, turf management is a billion-dollar industry and a specialist can always find work in parks, on golf courses, private homes, cemeteries and with landscape companies throughout the country.

Lawn Renovation Curriculum

- **Overview** of the turf care industry
- **Motorized tool safety** and operation
- **Lawncare skills**—planting, fertilizing and over-hauling
- **Direct management** of a facility's recreational and perimeter properties
- **Training** in renovating ball fields, public parks and cemeteries

SOURCE: CURRICULUM PROVIDED BY THE NCCI EDUCATIONAL HORTICULTURE PROGRAM IN GARDNER, MA.

Nursery

A prison nursery provides a cost-effective way to generate a source of planting materials that can be used for the prison or distributed to the community. The nursery can include planted stock, propagated cuttings in different stages of growth or full-grown shrubs, trees and perennials that are donated by nurseries or gardens.

At Rikers Island, the program has established relationships with different nurseries in Westchester County and Long Island. In late fall, after planting season, the nurseries donate all their unwanted nursery stock to Rikers. Plants are wheeled into a 50-by-50-foot area, mulched and left all winter into the spring when they are planted out, either in New York City or on Rikers. Afterward, the nursery is turned over and converted into a vegetable garden. By November, when the city's nurseries are closing shop, the site is again prepared for leftover planting stock.

A curriculum can be tailored specifically for establishing productive nurseries and include topics such as propagation (cuttings), plant classification, integrated pest management, pruning, root care, fertilizing, watering and transplanting. Marketing and pricing can also be included, to give students a full-scale experience in running a nursery. Have plenty of catalogs on hand to help students price out retail and wholesale figures for determining their inventory's overall value.

Floral Design

Flower arranging or floral design is more than a creative medium to express beauty through cut flowers, leaves and stems; it is also a billion-dollar nation-wide industry. Floral design helps students learn about flowers, plant growth, and plant names; identifying different shapes in flowers, leaves, and stems helps them refine their eye for texture and structure. Like art, floral design has therapeutic value, providing a sense of accomplishment and self-esteem while drawing students closer to nature. Occasionally we have formal workshops with professional floral designers who teach more traditional methods of flower arranging and the steps needed to establish a professional skill in the marketplace. Typically, however, students use the materials produced from a morning of summer deadheading, pruning and trimming out foliage to experiment with and create their own style.

Plastic vases, pots, and other containers for holding the

arrangements can be purchased quite cheaply. A few simple demonstrations of how to properly cut and prepare the stems and foliage and students are ready to begin their arrangements. Students must learn which plants bloom after deadheading, or where and how far to cut back roses or stems and branches with foliage, which flowers to leave, and how to use the garden in a balanced way without creating a discernible impact on the system.

Rikers Island: Living Color

THE MEN ARE DRIFTING. They drop their tools and wandered away from the greenhouse, the carpentry shed and the garden. Even Hariberto leaves the kitchen to stand with them in a line by the classroom to watch, silently, the most important moment in the day for them—the parade of female inmates leaving their program housing units for lunch in the jail mess hall. The women walk along the fenced-in corridor that borders the length of the grounds, almost 200 detainees in civilian clothes, enrolled in a drug program and housed in free-standing bubbles that resemble indoor tennis courts some 100 yards from the women's jail. Some of them wave, a few shout annoyances like "You my baby's daddy?" or "You know Romeo from the street? If he's in 76 tell him to write me!"

The men have been warned by Officer Periera: No talking to the women. Do not meet them by the fence line. Do not pass cigarettes or other things through the fence. So they stand there frozen. Occasionally they wave, or smile or mouth a silent phrase. In the past, we had men hanging on the fence line, women flashing their breasts, both groups passing cigarettes, notes and commissary food between the gate or through the fence. It was out of control and counselors from the drug program complained about the men harassing women on their way to lunch or commissary (though in actuality they harassed each other).

Pereira stepped in with a one-chance ultimatum. "Don't be stupid," he tells them. "I don't want to write anyone up, because it's more work for me, but I will if I have too. One word to them and you're out."

The women deliberately and slowly shuffle past them. Without fail, one of the guys sees someone he knows from the street.

"Yo, I used to sell to her on 170th. They popped her good, Yo." Or, "that shorty in the tights, Yo, was Junior's girl. He don't even know she's in here." And the all too common, "Yo, that's my cousin, Yo." Recognition for even the most casual acquaintance brings a retinue of smiles and waves and promises to write or pass a message to friends. There's a small-town feel about it, the flash of familiarity, the imparting of news about people they have in common, a quick update on their case or sentence.

"I got 10 days and a wake-up . . . "You lookin' good . . . "I'll see you on the street . . . "Tell Mousy to write me."

"The problem," Periera would tell me, "is too many people are too comfortable with jail. It's like a vacation resort they keep coming to year after year. They catch up with family, with neighbors with the people they deal with on the streets in their neighborhood. Shit, for a lot of these people, jail is an extension of their neighborhood. What does that tell you?"

One of the women slips a note through the fence and waves at Jason. A few of the women hang on the fence in a knot, as if observing the garden. They look absently at the plants in a distracted conversation. The men stare. The women, aware they have given ample time to display their goods, move on. When the corridor is clear, Jason heads to the fence to retrieve his note, risking the consequences

Periera comes out from the classroom. "Tools away," he says. "We're out of here in five minutes." Supplies and equipment go into the shed. Wheelbarrows are turned over and bags of garbage are hauled to the gate. Tyrone heads to the compost bins with organic peelings of the day's lunch. I also give him the ritual task of filling the empty bird feeders with food. "Feeding birds is good for your soul," I tell him.

I give a quick glance at the grounds surrounding the greenhouse ticking off our accomplishments and creating a mental checklist for the week ahead. The ground at the gazebo site is completely turned over and raked out. On the following Monday truckloads of conifer trees and evergreen shrubs from the winter planting at Rockefeller Center will arrive, some of which will form the backbone of our pond and gazebo garden. The path is carved out and ready for the sand and

The contrast between fence-lines: unchecked weedy growth and a newly cleared space for gardening.

stone. John Cannizzo with two new students have started building La Zona Verde. The summer flowering shrubs still need to be pruned. And the vegetable and herb gardens need to be prepped for the seedlings in the greenhouse. I have two new male students. Two others went home the previous week and another has transferred to a detail in laundry. Four women are in their first week, two women have left and two are at follow-up with the Women's Prison Association (WPA) as part of their impending release.

Attendance has been anything but uniform. The projects have been random and dictated by the weather and season. Often we have left the greenhouse by bus and traveled to different sites around the island where gardens are built that demand maintenance. Yet, for those who come each day, the greenhouse is a surrogate home, a place physically and perhaps spiritually removed from the calculated regiment of jail. Inmates find their way through the labyrinth of available projects and, in assuming responsibilities to build and create, develop their place in the garden.

The men head to the gate where they'll board the bus to be driven back to the sentenced men's jail on the other side of the island. "It's a shame we gotta leave so early," Hector says.

"I could stay here all day," Hariberto says. "Hey Periera, why can't we stay until it gets dark?"

Pereira looks at his watch. "Because at precisely three my shift ends, and, unless I'm paid time and a half, I will not be on this island one minute longer."

"I guess we're going back to jail," says Hariberto.

We head out the gate. "There's always tomorrow," I tell him.

CHAPTER EIGHT

Aftercare: Breaking the Cycle

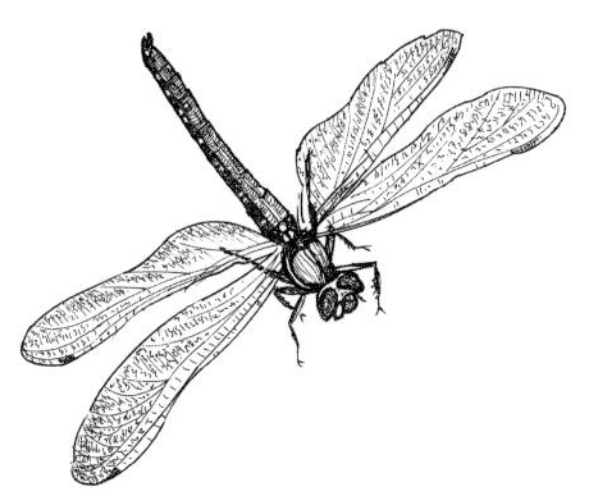

As of the end of 2002, the American penal system consisted of 5,033 adult prisons and jails.

—JOAN PETERSILIA, AUTHOR OF "WHEN PRISONERS COME HOME"

In 2003 the cost of the criminal justice system was just over $49 billion. That same year, the federal Department of Education budget was a tad over $42 billion.

—ALAN ELSNER, AUTHOR OF "GATES OF INJUSTICE"

Repeat Offenders

While the number of violent crimes committed in the U.S. has dropped considerably the past decade, the number of incarcerated U.S. residents continues to rise dramatically. In 1972 there were approximately 200,000 inmates in the U.S. By January 2004, there were over two million men and women adults and adolescents behind bars in the U.S., a per capita growth of 700%, and what constitutes a veritable nation of prisoners. A majority of those incarcerated were return visitors, part of the national recidivism rate of 67%, and one of the telling problems of a criminal justice system that finds success warehousing prisoners rather than reforming them.

In her book, "When Prisoners Come Home," Joan Petersilia states that just "one-third of all prisoners released will have received vocational or educational training." And with three-quarters of all inmates diagnosed with problems of drug or alcohol abuse, only one-fourth will be involved with some kind of program prior to release. Meanwhile, most of the programs offered in prison have long waiting lists. When they are released—more than 600,000 inmates will leave prison this year—they depart jail without the necessary tools to negotiate their lives in a productive manner. In California alone, of 132,000 inmates released in 2002, just 8,000 received some kind of aftercare support to help them successfully reenter and remain in their community.

Numerous studies have been carried out to determine the factors that are most effective in helping ex-offenders stay out of jail: inmates with a college education followed by a marketable skill are less likely to return to jail than inmates with no education or professional training. Yet, even with skills or a degree, ex-offenders often arrive home to conditions which defy the best intentions to remain straight or free from the parole violations that will land them back in jail. A spouse or close family member who is still using drugs, pressure to provide for young children or elderly parents, a good deal of idle time and the prospect of fast money in the drug trade add to the likelihood of an all-too-sudden relapse.

"It's like this," a former student and ex-addict told me. "No one wants to see you straight. They're all willing to give you free what they're copping on the street. Then, as soon as you're hooked no one wants anything to do with you. Don't even look at them if you don't have cash in your pocket."

Parole violators are especially vulnerable. According to the Bureau of Justice Statistics, two-thirds of those released on parole were back in jail within three years. In California, 60% of new admissions were parole violators, and four out of five of these were not convicted of new crimes but had technical violations of their parole. A violation can occur simply for leaving the state without permission, failing to report to counseling, failure in finding a job, or having contact with other felons.

Mosaic paver made by a student in winter embellishes a pathway.

In fact, in the garden the students speak all the time about turning their life around, getting clean, taking care of their family, and holding a job. But as a former student and repeat offender told me: "For most of us, change isn't just staying off drugs. It means leaving the neighborhood, leaving the husband, finding new friends, new family, finding a job and a program, leaving the old life completely behind. I'm not saying it can't be done. I'm just saying it's a lot harder than simply saying I need to change."

With the odds stacked against them, it's no wonder why ex-offenders—especially those arrested for drug related crimes—keep returning to jail or prison. But studies in criminal justice and recidivism agree on one crucial fact: ex-offenders who have jobs, attend programs or school, or receive vocational training shortly after leaving jail have a lower rate of recidivism—28% compared to the national average of 67%. In cost alone the disproportion between re-incarceration and rehabilitation—roughly $57,000 to $39,000 or less (some vocational education had a price tag of $2,000 per inmate)—is enough to make the taxpayer wonder why more funding isn't allocated from the criminal justice system into aftercare programs.

Aftercare not only helps ex-offenders find jobs or training, but it also provides them with short or long-term support as they navigate their lives on the "outside."

Rikers Island: Across the Bridge and Back

THE CALL, when it comes, is never surprising. Sometimes it is a girlfriend, or a husband or wife. And occasionally it is an employer who hired one of our students after their release. This one, at 1:00 AM starts with the familiar refrain, "Have you seen or heard from ______ (fill in the blank)?" And then the name sinks in. "Hi, this is Nancy. John's girlfriend. Have you heard from him?"

John P. is the carpenter who built the gazebo. Released and out for over a year, he was a program success story—an inmate who used his time in jail to construct a life that would keep him off the island. His private bridge off Rikers, he liked to say. Shortly before leaving jail, I hooked him up with a friend of mine who had just purchased a two-family house in Cambridge, Massachussetts, that needed work. John, who for the past decade had used his time out of jail for scoring crack in his Spanish Harlem neighborhood, thought leaving New York would break him from familiar terrain and his bad habits.

On a cold winter morning at 5 AM, a bus pulled up to a stop beneath the elevated subway tracks at Queens Plaza and the 59th Street Bridge and disgorged its entire load—released men from Rikers Island. By day, the spot is one of the busiest corners in the city, with subways converging for Manhattan and Brooklyn from all over Queens. By night it's deserted except for the customers in the Dunkin' Donuts and a line of cars leaving Scandals, the strip joint across the street. A few of the inmates were greeted by family or friends, but most left the bus alone, met by the Indian woman selling coffee and rolls from a cart, a few prostitutes and a couple of drug dealers skulking in the shadows of the elevated tracks. For many ex-inmates, the decision they make in the moments after leaving the bus, will determine the immediate course their life will take as they navigate their freedom. A majority makes a choice that will lead them back to jail within six to twelve months of their release.

John stepped off the bus, skirted the dealers, gave a passing glance to the crack whores and using the allotted $10 given to him by Correction when he left the island, bought a cup of coffee and a subway token. By 6 AM he was at my apartment on the Lower East Side. An hour later he was picked up, by a friend who needed help with a remodel, and driven to Boston with nothing more than a paper bag of his belongings.

He was given a room with a small kitchen, tools, $450 per week, a small organ to play music and a two-family house to remodel. Over the next six months, John had a steady gig of hard work as he ripped out walls, replaced beams, installed pipes, hung sheet rock, built windows and doors and painted. By summer he was asked to work on additional contracts around Boston. He received his Massachusetts drivers license, bought a pick-up truck, tools, and rented an apartment. He was studying for his state contractor's license and by all accounts, apart from occasional bouts of heavy drinking in Cambridge's last remaining working class bars, was free of drugs. John had turned his life around.

One fall day he called me on the phone to save a date in March. "I'm getting married," he told me. "I met someone, and I bought her a car. She's crazy about me." I congratulated him. "I owe it all to you," he told me.

The pagoda garden in fall and spring. Built in the aftermath of 9/11, the pagoda has a pathway studded with mosaic pavers.

"Not at all," I said. "You have skills, the drive, the will to make things happen for you. You did it yourself." Though I did think our time in the garden, the discussions we had that revolved around changing his destructive habits, and using opportunities to build a new life had something to do with it. Like a father watching his child take its first hesitant steps in life, I took immeasurable pride in my 49-year-old ex-addict's accomplishments.

The call from John's girlfriend Nancy came shortly afterward. "He's gone," she told me. "I think he's in New York. He's taken all my money, my jewelry, emptied our account. He's in his old neighborhood. I know he's back on crack."

"I'm sorry to bother you, but if you hear from him," she said, "tell him to call me." I put a call in to the local hospitals and police precincts and ran his name through the system at Rikers to see whether he was re-incarcerated. Nothing.

It was just a matter of time, I thought. And sure enough, on a March day, Officer Periera arrived at the greenhouse with the news that we had our carpenter back.

The timing was propitious. We were trying to build a peace pagoda that two students had designed as a memorial for 9/11. There were flaws in the design, and the structure thus far seemed unwieldy rather than elegant. John's return meant the pagoda would actually be completed, and the inmates he worked with would learn from a professional. This time his bid was a short one: two months for stealing cigarettes.

"You need counseling," I told him. "You can't do this alone." Over the past few months, in speaking with Nancy and my friends in Cambridge, a composite of John's conditions emerged: Alcoholism, domestic violence, substance abuse, anger, and depression, followed by intense mood swings. I was shocked. He seemed so functional in jail. I would have never recommended him to work at someone's residence knowing how fragile and potentially destructive he was. Now he was leaving Rikers the next day and taking a bus to his brother's in upstate New York to build an addition onto his house.

"I already talked to my brother," he said. "There's a drug program I'll be joining the town over. I'm going to get my pick-up back, my drivers license, and with the money he's gonna pay me I'm gonna get tools. Within a month I'm back in business. Then I'm gonna pay Nancy back the money I took. She's already dropped the charges against me in Cambridge."

"You need more than a drug rehab program," I told him. "It's not about drugs. You need to address the bigger issues. Maybe get some medical help for your depression. You need a clinical doctor or program and long-term work—on yourself."

"I'll be OK," he said. "Look, I'm getting too old for this. When I'm working nothin' seems to bother me. I just have to stay busy. In fact," he said, "if it wasn't for Nancy feeding my urge to smoke crack I wouldn't even be here."

A classic phrase, I thought. Blame the ex-addict girlfriend, and play the victim. He wasn't ready to take the steps to undergo the internal work he needed to change. Not yet. "Remember this conversation," I told him. "Don't waste your skills building things in jail."

John stuck his hand out for me to shake it. " I appreciate everything you've done for me," he said. "But James, I'm not coming back here again." He seemed both insouciant and determined. I wished him luck.

Breaking the Cycle: Post-Release Jobs and Services

For most ex-offenders, receiving a skill in jail and even finding a job once they are out is no guarantee they will readily adjust to life on the outside. Many inmates have deep-seated emotional problems, a history of substance abuse, and ongoing financial or housing issues. Having the skills to find a job does not mean they are necessarily able to hold one.

A successful aftercare program helps ex-offenders address their immediate needs after leaving jail. It may help an individual find transitional housing, work, or drug programs. It may provide legal aid, and offer help with school or training. But, to be truly effective, aftercare should continue to work with the client long after they become independent or begin to maintain a crime-free life. Through follow-up services, programs often require staff to have periodic discussions with their clients' employers regarding the individual's job performance. If their work is less than satisfactory, the program staff will intervene in an effort to help the client become

The garden and pathway wind past the greenhouse contrasting sharply to the fenced-in buffer adjacent to the women's jail.

a more valuable employee. If the client seems to have relapsed, aftercare can find them treatment before the problem escalates. This intervention is often the crucial difference between holding a job and losing one, between staying free and ending up back in jail.

Rikers GreenHouse

At the Rikers GreenHouse Program, there is no prescribed or single discharge plan for departing inmates. Instead, we work on an individual basis to help evaluate and determine each student's needs and interests. Does the student have a history of substance abuse and require in out-patient programs? Do they have a place to stay when they return to the streets? Are they married, do they have children or are they the sole provider of a family? Do they have an employable skill or career? Do they require full or part-time jobs when they leave Rikers? What is the kind of work they are interested in finding? What are their short and long-term goals?

An inmate's life is generally so fractured after they leave jail that handling work or school full-time can be overwhelming. Ex-offenders are likely to have issues in their personal relationships or with social services; they may lack stable housing; need appropriate identification; have meetings with parole officers or drug counselors; as well as a host of other issues and requirements that take precedence over finding and holding a job.

These profiles assist us in understanding the appropriate strategy for release. In general, if an inmate is under thirty, single, with a high school diploma or GED, unskilled with no children and lives at home with a parent, we may recommend full-time school. La Guardia Community College's inmate release program, called CUNY-Catch, offers tuition and loans to ex-offenders as well as remedial educational courses to help them obtain a GED and prepare for college. Computers, accounting and design are just some of the courses students can take to make them more employable for well-paying and interesting careers.

An inmate with children and a history of substance abuse with little experience in the work force and no financial support may need a program that offers drug counseling, job skill development and part-time work. Another inmate who is not interested in school and does not have marketable skills may require a vocational and job placement service.

The needs and interests vary, but for an in-house aftercare program to be effective, it should start in the garden program working one-on-one with students well before their exit date. At Rikers, we treat this as a process. As soon as inmates begin the program we enter their background information into our computerized database in the classroom. The database also assists us later in tracking the development and post release status of each student.

Knowing the release date, the level of education, work experience and future career or educational interests allows us to prepare an individualized release plan. During their time at the GreenHouse, while the staff is making the appropriate contacts with prospective programs, agencies and employers, the students are conducting research on jobs, programs, or housing, and they are outlining goals and developing resumes.

We are helped in our work by various programs on Rikers such as the Women's Prison Association (WPA), which brokers community-based or institutional services for women inmates leaving jail. The Rikers Island Discharge Enhancement (RIDE) program offers sentenced inmates pre- and post- release planning and services including drug rehab, recovery and transitional housing, as well as referrals to CUNY-Catch, Center for Career Employment Opportunity (CEO) and Osborne Association for continued education or job placement services.

Database and Tracking

We found the following categories to be most effective in developing profiles of our students for their exit strategy as well as for tracking their progress after their release.

- **Name of student**
- **Date of program entry**
- **Release date**
- **Phone**
- **Contact address**
- **Number of children**
- **Education**
- **Employment history**
- **Future interests**
- **License/certificates/language/jobskills**
- **Post release 3 months; 6 months; 1 year; 3 years.**

Tracking can provide data to determine the success of a program in reducing recidivism.

GreenTeam

At Rikers, most of our students are only at the GreenHouse for six to twelve months, too short a time for them to develop into year-round professional horticulturists. To supplement their learning and prepare them for careers, we offer released GreenHouse students an opportunity to work on horticultural projects in New York City as interns through HSNY's GreenTeam and Fee-for-Service program. The GreenTeam builds, installs and maintains gardens and conducts workshops for public/private agencies and individuals. Interns earn $7.50 to $10 per hour for their work (*see Chapter Nine*).

Working for HSNY gives interns a work history and professional skills they are able to use at their next job. They are also given time to attend drug programs, GED classes, or handle issues with housing, social services and their personal health. They learn to prepare resumés, and they attend classes and horticultural workshops.

Where to Start?

Inmates on Rikers have access to two important publications, which specifically address post-release programs, housing and social service agencies and employment services for the ex-offender.

Connections is a comprehensive resource guide published by the New York Public Library and printed at Rikers Island. Free to all inmates and ex-offenders, it offers such information as how to register to vote, how to restore your rights by cleaning up a rap sheet, writing a resume, and, for NYC residents, a borough-by-borough listing of substance abuse programs, transitional housing services, and educational and vocational programs available to ex-offenders.

Connections is also available on the Internet. With this valuable resource, a New York State jail program is able to research and offer substantial options and information for departing inmates. It is the cornerstone of any at-risk program in our state. Similar resources are available on a state-by-state basis through state parole and correctional offices. Or, if an effective regional directory is not available it may be feasible to compile and publish a "release survival guide" for departing inmates, or for the population the program is serving.

Programs for Employment

A large urban center such as New York City has a number of non-profit agencies that deal specifically with employment training and placement for ex-offenders.

CEO (Center for Employment Opportunity) for example, offers ex-offenders transitional employment the day they leave Rikers Island. Individuals earn $36 a day plus carfare for cleaning courthouses or public schools, painting public buildings, providing outdoor maintenance or demolition services. While the pay seems minimal, the service CEO provides is vital for it enables individuals, with or without skills, to begin work immediately and to develop a foundation for job readiness and work experience.

Employment lasts for six weeks, after which CEO provides job counseling and long-term placement with agencies, businesses and private contractors. They also have continuous follow-up with clients up to three years after joining CEO. Their record speaks for itself. Each year, over 3,000 ex-offenders receive transitional work; roughly half last the entire six weeks. Considering that CEO accepts ex-offenders without regards to their criminal record, work experience or skill level, the 50% retention rate is remarkable. More impressive is that 800 to 900 graduates are placed in jobs annually.

For Tim Williams, CEO's Director of Transitional Employment Services, the serious challenge is working with inmates who leave Rikers and shuttle beween shelters and jail. "There's a lack of stability, so a lot of what we do is try to keep people engaged who do not have community support," says Williams, "and if we lose someone with a drug problem or they disappear from work, then we re-engage."

At CEO, job coaching plays a major role in getting ex-offenders job ready and is instrumental in developing and maintaining relationships with the clients. According to Williams, this relationship helps ex-offenders stay focused as they pursue their goal of finding a job and creating stability in their life.

There are also job placement agencies such as the Fortune Society and the Osborne Association's South Forty Corporation. Unlike CEO, these agencies do not provide ex-offenders with work immediately, but require them to go through a job readiness and preparedness program before they are sent out on interviews for prospective jobs. South Forty also has work crews that, similar to CEO, have contracts to maintain public spaces through the city or state.

Some programs are more successful than others or have specific criteria for applicants. Because one bad experience can sour a student's interest and hope in the job search, it is the staff's responsibility to research the track record of each program—the types of jobs they offer and their rate of placement before referring an inmate to a program.

People in prison can often offer helpful information about services, and drug and job programs they have been to in the past or heard about through other inmates. It is important to personally contact program personnel who have a direct line for jobs or training, or who can work with students after their release.

Program staff can also tap into TRAIN (Training Resource and Information Network), a computer database that has thou-

Profile on the Tree Corps

Planting and maintaining street trees was one way for released inmates from the San Francisco County Jail to earn a living while improving the quality of life for San Francisco residents. The Tree Corps was the brainchild of Cathrine Sneed, founder of the city-based Garden Project, a program that builds on horticulture, farming and nutrition to keep ex-offenders from returning to jail. Ms. Sneed saw an opportunity to provide former inmates with tree work during her stint with the Citizens Advisory Tree Commission. Money set aside for transportation initiatives was slated for landscaping and planting trees on transportation land. And, in 1992, the Tree Corps began when Sneed's gardening group was offered a contract to plant 800 trees.

The native woodland at our Rikers garden allows students to familiarize themselves with trees and the opportunities on the outside to find work as tree care professionals.

The following year the San Francisco Department of Public Works extended the contract to plant and maintain trees for a period of two to three years. The agency threw in four DPW trucks, and, with ten former inmates, the program was off and running. Corps members were paid an hourly wage of $11, along with health benefits, and worked closely with a counselor to address other issues, such as drug abuse, housing and obtaining a GED. At the same time, the program also helped Tree Corps participants move on to other avenues of employment.

By year 2000, up to forty people worked for Tree Corps. Charlotte M., who was a long time drug dealer and single illiterate mother, showed little interest in finding work or new skills that could make her employable after her release from jail. By chance she hooked up with Sneed's Garden Project and began growing vegetables and herbs to fill time and supplement her food income. She then joined The Tree Corps, and its focus on both work and school gave her the impetus to learn to read and write. She left Tree Corps and started her own business, a day care center, and has lived the past few years as a successful homeowner and entrepreneur.

"Life is good when you set goals," said Lance T., a current Tree Corps member and former drug dealer. "Now when I'm outside I'm planting trees or passing out vegetables instead of looking over my shoulder." ■

*Tree Corps was discontinued when the program lost its funding from the City. It remains, however, a valuable model for providing green jobs to ex-offenders and at-risk individuals.

sands of listings of job and job training resources as well as childcare and substance abuse treatment in the New York City area. TRAIN is free for anyone with a valid New York Public Library card. Otherwise, TRAIN charges for the service.

Careers in Horticulture

With their training at the GreenHouse, a number of inmates leave Rikers with a demonstrated interest in long-term work in horticulture. The GreenHouse staff works hard to generate relationships with local professionals who can help students with potential jobs. Tree care companies, nurseries, private landscapers, interior plant care firms, greenhouses, golf courses, botanical gardens, zoos and city parks departments are all viable options for employment. As in any field, the more expertise the student brings to the job, the better opportunities they have for advancement and higher pay.

There are also a number of professional schools that provide courses and certificate degrees in horticulture. The New York Botanical Garden offers programs and professional degrees in gardening, horticulture, horticultural therapy and landscape design. The Brooklyn Botanic Garden provides courses in gardening and composting, and the Horticultural Society of New York has adult education classes and a certificate course in floral design. Trees New York offers a certificate citizen pruning and tree-climbing course and Bronx Community College has an associate degree in horticulture.

Some major tree care companies, such as Bartletts and Asplundh, offer internships to learn pruning, tree climbing and chipping. These are year-round jobs that offer skills, good pay and benefits and offer opportunities to work all over the state or region.

Program staff need to search and locate the larger companies in each region or area and inquire about their policy on training and hiring interns. Metro Hort, an organization of horticultural professionals in the New York area, offers a website with job postings of horticulture positions and internships all over the metro area. We emphasize to all our students the

Georgia Justice Project

The Georgia Justice Project (GJP) is an eclectic mix of lawyers, social workers and gardeners who for 20 years have defended the poor accused of crimes and, win or lose, helped them rebuild their lives. Founded in 1986, GJP's mission is to ensure justice for the accused and to take holistic approaches in assisting them to become productive citizens. In order to provide clients with jobs after their release, the project employs them in their in-house landscape company—New Horizon Landscaping.

GJP carefully selects which cases it takes on. People are initially referred to GJP because they have a criminal case pending and cannot afford an attorney. They become GJP clients if they are committed to making a life change during their legal course work.

The client's initial contact with GJP includes both a legal and social work assessment. A holistic case plan is developed as a team. Services include quality legal representation, individual counseling, substance abuse intervention, educational assistance, and job training and placement. If clients are convicted, GJP visits them in prison, advocates for their needs while in prison and provides post-release support.

To ensure all clients have the opportunity to receive job training and steady work, GJP clients are eligible for pre-disposition and post-release employment in its company, New Horizon Landscaping (NHL). New Horizon Landscaping has provided lawn care—maintenance and installation—since 1993, servicing residential, commercial and industrial properties in Atlanta. Lawn care is in high demand in the expanding suburbs and commercial plazas of Atlanta, and this simple but elegant business is critical in helping the indigent accused change their lives. ■

For more information see www.gjp.org

importance of seasonal internships with the city's major botanical gardens, parks and zoos. This not only builds professional experience away from GreenHouse, but internships can potentially lead to full-time jobs with that particular organization.

Outside New York City demand is growing for workers with experience in environmental restoration to repair the nation's damaged streams and riverbanks, coastal beaches, flood plains, wetlands and other ecosystems. In urban centers there is a steady stream of restoration projects around landfills, degraded waterfront areas, greenways, greenbelts and other public space in need of improvement. To tap this stream,

Spring tulips blooming in the garden are in stark contrast to a prison landscape of bare site lines.

program staff must become familiar with contractors or subcontractors working on each project, or create their own enterprise, similar to the GreenTeam, that can bid and manage their own contracts.

Building partnerships with existing non-profits is an essential step in establishing an effective after-care system. It allows a program to extend its boundaries past the walls and fence-lines of jail into the communities the students return to. Meanwhile, the partners benefit from inmates who leave jail prepared for a particular job or training.

Inmates associated with a vocational program in jail are more likely to accept guidance from program instructors than from a correction or probation exit counselor. At the same time, job training and placement programs, shelters, transitional homes and drug programs vary greatly in quality and effectiveness. Word travels fast in jail and on the street. It is important to ensure that any partners in aftercare have a well-established record in providing ex-offenders the services they need.

CHAPTER NINE

The GreenTeam: Growing Roots in the Community

For the first time in my life I have a sense of purpose when I wake up and go to work. I have the feeling that whatever I do is going to make someone happy and leave a site looking beautiful for other people to enjoy. I never thought making a garden could do all this. I feel blessed.

—EVELYN M., FORMER GREENTEAM INTERN

Perhaps the most satisfying moment in my day is to meet a student on a city street in New York wearing jeans, a t-shirt and a pair of boots—no tell-tale sign of Rikers prison orange—ready for their first day of work with the transitional employment and vocational training program we call GreenTeam. It is a disconcerting moment; the face so familiar the past six months at Rikers seems out of place with the roar of traffic and the high-rise buildings lining the sidewalk. No fence lines. No officers. No uniforms. Despite the personal trials associated with "coming home," the students, now interns, seem physically buoyant, unburdened by the stress carried in jail.

Part of the garden designed by John Cannizzo and built by the GreenTeam in the back of the Aguilar Branch Library on 110th Street in Manhattan.

As we meet, there is a quiet hope in both of us that what took place at the Rikers greenhouse—the desire to change, cope and make a productive life—will now play out in a city park in mid-town Manhattan; in a penthouse garden on the 14th floor of a luxury building; or in any of the spaces we get paid to install and maintain gardens. GreenTeam contracts cover a variety of grounds in New York City, including city parks, the NYC Housing Authority where we conduct children's gardening workshops, and private buildings and with over 15 libraries where HSNY installs and maintains learning gardens for the branch public library system. Most important, inmates leaving Rikers have the opportunity to find immediate work either full or part-time as they acclimatize to life outside of jail. One supervisor oversees the interns' work (typically we have up to six full and part-time interns on the "team") and is responsible for acquiring clients, maintaining contracts and designing projects.

The contract between us is clear: 1) use what was learned at the greenhouse; 2) be willing to learn and take orders; 3) come to work on time; 4) exhibit professional work manners on the job-site; and 5) stay clean, and we will do what we can to move you into a permanent job working with public or private agencies, organizations and businesses in the field of horticulture. We pay a beginning wage of $7.50 per hour, which can go as high as $10 depending on the intern's level of skill and responsibility they assume on the job site. This is the "street" part of our "jail-to-street" program, and how the intern performs over the next nine months determines, for the most part, our effectiveness in keeping ex-offenders from returning to jail. Woody Allen once remarked that 80% of success is just showing up. For interns, a few days removed from being incarcerated, showing up, or calling if they cannot work that day, simple as it sounds, is harder than assimilating many of the skills and knowledge they need to learn for a full-time job in horticulture.

Carol D. had been out of Rikers for almost a week when she called us for work. John Cannizzo, now the GreenTeam Director, already had a crew of three working at the 79th Street Boat Basin where we had a maintenance contract for the gardens by the restaurant, so Carol met me at a site on 67th and Park Avenue. The GreenTeam was ideal for someone like Carol. For five months she had re-built, designed, planted and maintained the herb garden at Rikers, and had demonstrated on a consistent basis a commitment to her work and the garden. She had spent years playing money scams and con games to support a drug habit.

"I would play to people's greed," she told me. "Show them a little money, and no matter how crazy the story sounds, once you promise more—that's when you got them."

Carol put her two daughters through college while she plied herself with drugs and fought off two relationships with abusive men. At 50 years old she bore an uncanny resemblance to Whoopi Goldberg. She was ready to change her life and create a semblance of stability with both a job and a place to live. The garden at Rikers helped her focus, she told me. It was the one activity that cleared her mind and let her effectively evaluate her life.

"I realized that I need to control my environment better than I have," she told me. Perhaps that was an understatement. Prior to the arrest that brought her to Rikers, Carol shared a four room apartment in Harlem with a full-grown Bengal tiger named Ming. When Ming took a bite of Carol's roommate's leg (the tiger's owner), Carol alerted the Bronx Zoo, which

a backyard in Brooklyn. For two years, the GreenTeam worked with children in a community center at housing projects in Harlem and the Lower East Side, building vegetable gardens, planting bulbs and sprouting plants from seeds. Every Tuesday for one school year, John and two interns also helped prepare and teach a greenhouse curriculum with special education students at Frederick Douglass High School in Harlem.

After some time, the GreenTeam and its schedules, projects and interns, appeared as a shifting mosaic of people and plants and human drama played out on various stages throughout the city. It was not just a question of finding the contracts, but a matter of working with interns through complex issues concerning housing, family, children, health, the courts, and the looming specter of most former substance abusers—a relapse and a return to jail. The idea, of course, was to give our interns skills and enough support to maneuver past those hazards into a full-time job somewhere else. And the hazards were everywhere: A new drug addicted boyfriend attracted to the prospect of a working girlfriend and her bi-monthly check; a mother dying of cancer in another state and the prospect of parole violations; a child in foster care arrested for sexual harassment; a drunken spree which leads to a fight; the breach of a court order of protection; the discovery that a partner is HIV positive; the loss of an apartment; a brother selling drugs; and the constant need for quick money. One of these or more would cause an intern to miss work without calling, or to simply disappear for awhile—from a couple of days to as long as a year. But the work would go on. At any time, we could always count on one or more interns who would serve as a core of stability, who knew the clients, the work and had the prospect of moving on to something better.

Now, less than 5 days after her release from Rikers, Carol and I ride an elevator in a small luxury building that takes us directly to the penthouse. I am carrying a canvas bag with a

The Outlook for Hiring in the Horticultural Field

Per Bureau of Labor Statistics (BLS)

BLS identifies the skills, tasks and responsibilities necessary for the successful performance of landscape workers' jobs as the ability to transplant, plant, mulch, fertilize, and water flowering plants, trees, and shrubs, and to mow and water lawns at residential and commercial sites such as shopping malls, multi-unit residential buildings, and hotels and motels on a regular basis during the growing season. Other sites identified by BLS include golf courses, cemeteries, campuses, and parks. According to the BLS:

- Landscape workers held about 1.1 million jobs in 2004 of a total of about 138,000,000 jobs in the total employment of America (about 1 out of every 138 workers).
- More than 1 out of every 7 landscape workers is self-employed.
- Most workers have HS education or less.
- Short term on-the-job training is common.
- Most sought after qualities are responsibility, self-motivation and ability to get along with others.
- Those interested in maintenance occupations should find plentiful job opportunities in the future. Employment of grounds maintenance workers is expected to grow faster than the national average for other occupations due to increased construction of new housing and emphasis placed on greening commercial spaces such as malls, sports complexes, highways and recreational facilities, as well as upkeep and renovation of existing spaces.
- Driver's licenses are very important for employment.
- First-line managers can earn $15 per hour but require some formal education. Median hourly salary for entry-level workers is $10 per hour.
- Landscape work is stable and likely to be increasing in demand by up to 35% over the next 10 years.

SOURCE: BUREAU OF LABOR STATISTICS: OCCUPATIONAL OUTLOOK HANDBOOK; GROUNDS MAINTENANCE WORKERS, 2004

few hand tools and plant fertilizer. This job is a quick visit to a stunning wrap-around garden on the 15th floor of a Park Avenue building. "Mostly a clean-up and feeding," I tell her.

Much of the garden contains a large display of healthy, well-grown conifers that form an archway along the terrace. Birdhouses and feeders sway from branches. A wisteria vine twists up the brick façade of the penthouse. A few less healthy birches and plums and two hophornbeams have out-grown their planters and are showing signs of dieback. The owner, an HSNY Board member who years back was instrumental in establishing a successful culinary and journalism program for incarcerated men called Fresh Start, does most of her own maintenance in the garden. It was Fresh Start that planted the seeds for our fee-for-service program with their own short-lived culinary business, "Catering with Conviction." The idea was elegant—students from the culinary class, released from Rikers, could work in a catering business run by Fresh Start, based in Manhattan. The business succeeded for a while, catering parties and events around the city. But it underbid its jobs and eventually folded.

Carol and I work quickly, cutting back dead branches, aerating the soil and adding a top dressing of manure to the planters. From the garden looking south a ragged jumble of domes, cornices, pinnacles and rooftops carry us to mid-town. The skyline is punctuated by patches of green; rooftop gardens that from here seem nothing more than epiphytic plants nestled in the leafless crowns of the towering buildings. I look at Carol. "As your own client, this would be a hundred dollar job," I tell her. "But for this we'll charge her thirty. She's a big supporter of our work."

In an hour we are back on the pedestrian domains of 67th Street heading to the Armory, home to the famous 7th Regiment National Guard.

The Armory with its peaked iron roof and brick facade takes up an entire city block, from Park Avenue to Lexington flanked by 67th and 66th Streets. When not in use by the Regiment, the Armory serves as a convention center of sorts for antique furniture shows, fabric displays, outdoor ornaments and other more obscure collectibles. Inside the drill shed the space is large enough to house a modern regiment and their associated hardware, which, as recently as 9/11, it did. The interior space, the place where meetings take place, where officers have lunch and where visitors first enter is a fantastic and surprising vision of deep, rich, carved wood, ornate staircases and heavy velvet drapes. Large portraits of past regiment members hang in gilded frames. Built in 1879, the Armory, after a struggle to replace it with luxury buildings, was protected when it received landmark status in 1994. On the second floor is a locally known, expensive restaurant.

A GreenTeam intern gets ready to plant a street tree in Manhattan.

"It's one of those places," I tell Carol, "that you have to already know. If you want to impress someone, take them there."

But approaching the Armory we have plans other than food, the 7th Regiment or an obscure furniture show. The Armory served an additional purpose; housed on the 3rd and 5th floors, in large open airless rooms crammed with beds, was a women's shelter. This was the hidden side of the Armory and for years an issue of contention with neighbors in the posh Park Avenue neighborhood.

Wrapped around the outside of the Armory behind a wrought iron fence was a landscaped area that once served as a complementary garden to the castle-like façade. Now however, the space was ignored and neglected. Apart from a few scraggly holly trees and a row of large London Planes, there was no vegetation; the soil was hard like baked clay, and the entire area, the length of 66th Street, around Park Avenue to 67th Street, was riddled with rat holes.

The site had the elements for a perfect GreenTeam experience: a beautiful landmarked building, a neglected garden that was an eyesore to the wealthy neighbors and homeless women residing in the building with time on their hands. All it took was a little seed money, permission from the Armory and shelter, and interest from the women residents to manage the garden, and we had a project. The money for plant materials and GreenTeam stipends was contributed by a long-time resident on 67th Street, and after an initial visit to the shelter several women expressed interest in learning to garden.

We focused on one part of 67th Street near the entrance to the shelter, where the women would congregate and smoke. By the end of one month we had already made a dramatic difference. Numerous bags of compost brought from Rikers Island was mixed into the soil. As a shady site, hidden by the building and shadowed overhead by London Plane trees, we opted for hardy shade tolerant shrubs and perennials that would offer a mix of texture, foliage and color through much of the year. With a rich addition of mulch, the site was transformed into a healthy shade garden. The response was overwhelming. The existing hollys seemed to thrive with the attention. Earthworms found a home under the layers of mulch and compost and their numbers exploded. The administration heading the Armory was thrilled, as was the Lenox Hill Association responsible for the shelter. The city came and poisoned rats. A core of three women kept the garden clean and watered, and the hundreds of residents who passed by gave daily encouragement. Or at least we took their comments for encouragement.

Of course there were problems. The women in the shelter wanted to get paid. They were often heavily medicated, which limited their work, and they often fought among themselves. The site demanded some kind of sustainable plan and funding, but until then, I visited weekly bringing a few plants, some mulch or compost, and slowly expanded the garden down 67th Street.

When we arrive, Lucy R., one of the shelter's steady gardeners, is waiting for us. Carol jumps right in. It would be a quick planting of New Guinea impatiens, an annual that could tolerate shade and add summer color. Lucy, who because of her weight has problems scaling the waist-high fence, leans on the rail and points to the area that needs flowers. A few women on their way to the shelter stop to smoke, adding their comments.

By 4 PM our day is finished. Lucy pulls the hose from the basement of the Armory to water the impatiens and the rest of the plantings. For her work Carol earns $7.50 an hour. She will not be paid for another week, so I advance her enough money to handle transportation and a week of food. In the past, interns received a monthly Metrocard (an unlimited transit card for the bus and subway) when they joined the GreenTeam. But with a possible 10 to 14 full and part-time interns, we could no longer underwrite the monthly transportation expense. Interns were now on their own getting to and from work.

For Carol, the money, or lack of money, is not an immediate issue. She is living with her boyfriend at his mother's apartment on the Upper West Side. Her children are grown and doing well.

"I'm the problem," Carol tells me. "Not my kids."

Despite the stability, she knows it is based on a shaky foundation. Her boyfriend is still using drugs, a major hazard for a former addict. And it is only a matter of time before the constraints of living with his family will get to her. She is also realistic.

"Give the internship nine months," I tell her. "Learn as much as you can. Take courses when they're offered," (Our organization has a lecture hall and offers a number of classes and courses taught by different horticulturist professionals through the year.) "Keep working, ask questions. Within a year you'll find a professional job in horticulture I can almost guarantee it."

If I sound optimistic, it is in part to encourage her to ride out the low wage of the internship, and also because I do feel confident that professional jobs are out there for those who work, stay straight and continue to develop their skills.

When I see Carol two weeks later she has already been planting trees at a new hotel on the Lower East Side, installed a garden on a rooftop near 62nd and 3rd Avenue, replanted an annual garden at a luxury building on East End Avenue and taught children how to plant corn at a housing project in Harlem.

A view of the Flatiron building as seen through the newly restored Madison Square Park.

This time we meet at Madison Square Park on 23rd Street in Manhattan. Ten years ago, Madison was just another neglected park the city could no longer care for. The emergence of the Madison Park Conservancy, a non-profit designed to fund, restore and help manage the park, began a dramatic transformation of a major gem in the chain of city park space. Depleted grass gave way to garden beds. The fountains were repaired and the homeless disappeared along with the drugs and dealers. Dogs were confined to an enclosed, self-maintained dog run. Flowering trees, shrubs, bulbs, perennials and ground cover emerged from a monotonous grove of viburnums, compact soil and rat holes. One million dollars was used to revamp the playground. For those like myself who recall the Park's sad state in the early 1990s, its renewal was nothing less than a spiritual revival.

It therefore seemed like an appropriate metaphor when GreenTeam was asked to help the head horticulturist Kim Wickers with a two-day planting of spring perennials. This led to a day of spreading mulch, and then a series of park-wide clean-ups, plantings and the installation of a playground garden. For over two years Kim has trained and hired a number of interns who have honed their skills in this park.

It is 8 AM when Carol arrives at Madison Square Park, a half-hour late. Lorenzo M., Marisol S., and Phil M. are already working spreading mulch. Today is Phil's last day with the GreenTeam. For the past month he has worked strictly in Madison Park; typically, Kim requests the most skilled intern to work with her, and Phil has skills. He also has a drug habit that he has battled off and on for the past two years I have known him in jail and with the GreenTeam. Off drugs, as a gardener, Phil is an asset to anyone who hires him. He is smart, charismatic, a hard worker and perfectionist in the garden. Unlike many of our interns or students, Phil comes from a solid, almost privileged background. His father was a well-known composer. His mother was a former elementary school principal. They lived in a beautiful brownstone on the Upper West Side near Harlem. Duke Ellington, Cole Porter, and Bobbie Short were contemporaries of his father and family friends.

An intern working in one of the perennial beds at Madison Square Park.

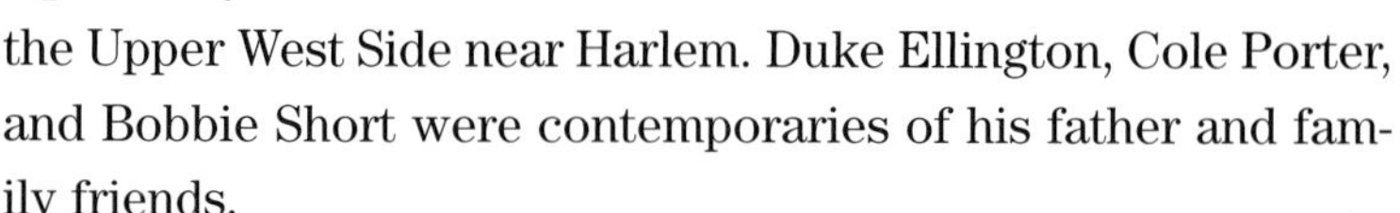

Phil went on to study fashion design at New York's Fashion Institute of Technology. The fast life of a struggling designer, he told me, was his downfall. "Too many parties," he said. "And too much drugs." And when Phil took drugs he had one annoying work habit—he simply disappeared. In one year this was his second stint with the GreenTeam. Having suddenly dropped out of sight for several months, his appearance was welcome, primarily because we seriously felt that Phil had the qualifications and skills to land a full-time professional job in horticulture. As long as he was straight and reliable we offered him a second chance with the GreenTeam. Second and third chances were an arbitrary decision based on the current size of our crew and the amount of work available, as well as the intern's past performance record. After all, although the GreenTeam operated as a competitive landscape firm, our clients knew we were providing learning opportunities to former inmates from Rikers. Any slip on our end could jeopardize future contracts with clients or potential employers for interns. We were far more scrutinized than other companies. On several occasions I had to appear before a building's board of directors to explain in depth what we did, the supervision we provided, the nature of crimes our interns committed (I would generally say drug-related non-violent), and answer any questions members would have regarding our program. I would often end a meeting by stating that any number of contractors entering the building had a criminal record. The residents just weren't aware of it.

"In our four years of operation," I would add, "we have never had a problem or issue with a client. How many New York contractors could make a statement like that and actually mean it?"

After one month assisting John with various projects around the city, and another working with Kim at Madison Square Park, Phil applied for and received a job with the Riverside Park Conservancy as a gardener making $15 per hour. He would work the summer and fall. After that, he told me, he was going back to school for fashion design.

"I love horticulture," he said. "It's something I'll use to fall back on if my career in design doesn't happen. I mean, that's a business you have no control over."

We were taking a break on a park bench waiting for Kim to deliver more mulch.

"This is one of the few jobs that people are really appreciative of what you do," Phil said. "You'd be surprised at the number of people that actually stop and thank me for the work I'm doing. No one will thank you for building a skyscraper, or driving a truck, or designing ladies wear. But they thank you for building a garden."

I couldn't agree more. And when I thought about the interns, gardening was their chance to have a profession that was openly admired by the public and something they could rely on for an income. All it took was coming to work each day and learning as much as they could about plants and gardening. It was the message John and I repeated to all the interns—work and learn.

I thought about the interns that had been with us the past four years. For those who returned jail—less than 10%—or disappeared, or never received a permanent job, it was not a question of skills that had failed them. Nor was it a lack of money. In each case it was the intern returning to a life on drugs. The GreenTeam was not going to be a 100% solution, but it gave our students something to believe in during their incarceration and it kept them involved after their release.

"Idle hands is the Devils playground," Phil would tell me. "I'd be back on drugs if it wasn't for work. I need to keep busy. Otherwise I go looking for trouble."

I thought of Erin O. She spent a couple of months at the greenhouse but quit after she had a disagreement with Officer Ross. Two months after she left Rikers she was re-arrested. When I saw her in the women's jail she was four months pregnant. Shortly after, she lost the baby, and once she regained her health joined us in the greenhouse where she spent the next six months serving out her sentence. Erin was 31 and Irish. Since the age of 13 she had been arrested, she claimed, over 150 times for prostitution and drug possession. At some point she married one of her clients and had two children, both of whom were raised by her sister.

I noticed right away that Erin did not take well to sitting through classes in the greenhouse. I surmised, correctly, that she had a learning disability; she was virtually illiterate. It was winter and I sent her into the garden and showed her how to prune the fruit trees and roses. I put her in charge of managing rows and rows of germinating seedlings in the greenhouse.

"I need a job when I get out of here," Erin would say over and over again. "I'm sick of my life. I need to change." She was leaving in April, a perfect time to be joining the GreenTeam.

"Call John Cannizzo when you get out," I told her. "Let him know when you're ready to work." She even had me telephone her father from Rikers, explaining who I was and the kind of work she'd be doing on the outside.

"My parents are watching out for me this time," she said. "But Jiler, I need work."

Erin was released on a Tuesday morning, driven off the island by the WPA (the Women's Prison Association) and by Thursday appeared for work with the GreenTeam. She quickly fit into a solid routine, appearing at John's apartment by 7:00 each morning with a coffee and roll, a pack of Newports and her lunch. At the end of the day, John would call her mother who would pick her up when she emerged from the subway at her stop in Canarsie.

Erin worked every weekday through the summer and fall. It was the longest legitimate job she ever held and the longest time in her adult life she had been drug free. As her skills developed, John would send her to work sites alone. She was dependable and reliable, two traits often lacking with interns who have spent much of their adult lives in and out of the system. More so, she enjoyed her work; I had the feeling it was for more than just a paycheck that she was out each day wrestling plants to the tops of buildings, or digging tree pits in the hard city earth.

By late December the workload had dried up and Erin had

few prospects to keep her employed through winter. Kim gave her several weeks of work cleaning the park, along with other miscellaneous chores. There were occasional pruning tasks at a luxury high rise on East End Avenue, a few construction projects that came John's way, and we put her to work in the Society's plant shop. She had to only hold until spring when the contracts picked up again. Meanwhile, John helped her develop a resumé, and showed her how to use the Internet to search for jobs.

In the time she left Rikers Island, Erin had established a legitimate background in urban gardening. She had an understanding of design, materials, soil and the names of major plants used in city gardens. She had work skills that could help her move to the next level of employment. By late February, Erin had a list of possible employers in the plant world. She had been out for almost a year, the work was picking up again, and then Chelsea Gardens, a prestigious Manhattan plant and landscape store, asked her to interview for a full-time position. As with all our interns, John prepared her for the interview, holding mock sessions that would cover her experience, the time she spent in jail, and the type of horticultural work she had professional skills in.

Chelsea Gardens offered her the job. She also received an opportunity to work in a union job at the newspaper. After 18 years of alternating her life between jail and the streets, of shattered relationships with her children and family and of abusive relationships in her personal life, Erin appeared heading for a happy ending. But, success may have been too difficult a prospect; perhaps it would negate the years of low self-worth she carried inside her. Or perhaps she just grew tired of fighting her addiction to drugs. A day after receiving word from Chelsea, Erin disappeared. A week later she resurfaced in her old haunts of East New York. She had relapsed. But this time Erin did something uncharacteristic of her past behavior. She checked into rehab.

We lost touch. Her mother and sister left Canarsie and moved to a house in New Jersey with no forwarding number. But wherever she is, it is clear that Erin has not been back to Rikers, and in that alone I see a measure of hope that her life has changed for the better.

When I think of Erin, I also think of interns who have used their experience to develop their careers and lives. Jackie A. occasionally calls John. She has been at the same job since she left the GreenTeam about three years ago. She has moved from her position as a floral arranger to an assistant supervisor to a supervisor/driver for Wal-Mart establishing horticultural displays all over the state. Clarence W., who had served time at age 17 was a member of the first GreenTeam and earned a BA in Finance from Baruch College while working as a job coordinator for a non-profit that provides services to the homeless. He is now working on his MBA. Lorenzo has been with us for a year. He is now an assistant horticulture instructor teaching mentally ill patients at a psychiatric hospital in Brooklyn.

Evelyn M., aka Fudgie, worked for us for over a year. She was 40 years old when she left jail. A long-time addict, she had never held a full-time job, and spent a good many years on park benches, in homeless shelters and on Rikers Island. It was February when she left jail, and with the cold and snow the GreenTeam was idle and not yet accepting new interns. She was a hard worker and during her time at the GreenHouse had shown a strong interest in horticulture. I told her to hold on until spring, April 1st at the latest. She joined CEO and immediately began work making $36 a day cleaning the courthouse in Brooklyn. The contract was long enough to keep her employed through winter, to keep her busy and away from drugs, and to instill in her the concept of waking up each day to a legitimate job. By the time she finished CEO, it was late March, spring work had begun and Fudgie joined the

GreenTeam, earning $7.50 an hour. She proved to be a reliable worker, and soon began alternating between training with John and working with Lisa Greenspan, the HSNY horticulturist in charge of maintaining the branch library gardens.

By mid-summer, Kim had specifically requested Fudgie to work with her at Madison Square Park as a GreenTeam intern. It took almost a year before Fudgie could ratchet her work up several notches to be considered a serious gardener. It required her to dress in a uniform, and to leave her personal life, her cell phone and her cigarettes out of the garden. It required her to have a continuous interest in learning, especially plant names. And most important, it required her to love the garden. None of it came easily. On several occasions she quit or was told to leave. But Fudgie would come back pleading for another chance to make it right. The gangster clothes disappeared. She carried a book of different plants practicing the names in Latin. She took a floral design certificate course and rooftop gardening classes. Kim noted the change. Subtle at first, and as the year progressed, more pronounced.

"Fudgie's indispensable," Kim told me, preparing to leave for a week in Berlin. "I couldn't be going away if it weren't for her."

In July 2005, the Madison Park Conservancy hired Fudgie as the assistant gardener at $15 per hour. For me, her new position was more than a culmination of hard work, desire and discipline that kept her on track and away from drugs despite a dramatic and often turbulent personal life. After years of neglect Fudgie personified the park's own transformation into a beloved city resource. Yet, to the rest of the world passing through the park on a warm summer day, she was simply a professional gardener.

When I see the list of interns who have worked with us over the years, I can't help but feel torn by the mix of those who have made it and those who were simply not ready to change their lives in a sustained manner. But I've also seen interns relapse, become incarcerated and then return to their work a bit more determined to succeed. I think of William, who despite an interest in establishing a career in gardening, initially appeared late to work or simply didn't show. It took months of stern warnings before he demonstrated a measure of reliability in the job place. One day after a week of punctuality followed by an unexplained absence, William simply said, "You may not see the changes, but I see them. I can tell I'm thinking and doing things differently."

William takes time to smell the roses in his community garden plot on 9th Street in Manhattan. For him, gardening has become more than a job.

True to his experience, the change became more evident as the year progressed. William entered a Citizen Tree Pruning course given by a local non-profit—Trees New York—in order to become a "Citizen Pruner." He also became a member of my local community garden on 9th St. in Manhattan, joining 64 other people in maintaining his own small garden plot. Here was an intern, less than one year removed from jail, gardening not for pay, but for the pleasure of it. The garden made him feel part of a something, he told me, even though he lived in his sister's apartment at the opposite end of Manhattan.

"A community," I said, qualifying his "something."

"Yeah," he said. "A community."

Working with the GreenTeam and assessing its impact on people's lives I keep coming to the same conclusion—change is a slow and often difficult process. But for those who work long enough in a garden to witness nature's renewal as it occurs on a seasonal basis, the capacity for self-renewal grows. At the least, transforming a bit of city landscape begins the process of helping ex-offenders transform themselves.

ACKNOWLEDGEMENTS

I WISH TO THANK a number of people who were instrumental in the establishment and support of our program on Rikers Island and in the creation of *Doing Time in the Garden*. Foremost, the program would not have come about without the insight and effort of Horticultural Society of New York President Anthony Smith, who parlayed his experience in city government and long-time interest in the criminal justice system to re-establish our place on Rikers. Arthur Sheppard, my predecessor at HSNY who first developed a horticulture program for jailed adolescents on the Island, initiated the GreenTeam, and helped me navigate the corridors of DOC. John Cannizzo, my co-instructor for several years, brought a wealth of ideas and talent, and now heads our *GreenTeam* Aftercare Program. A special thank you goes to Ms. Millicent Johnson who, as former HSNY Board Chair, was a source of inspiration and leadership during our first years on the Island. Kate Chura, Vice-President of HSNY, along with the staff of HSNY who make our small but dynamic non-profit a fun and accomplished workplace, have played a continuing role in the success of GreenHouse. Hilda Krus who, as the HSNY Horticultural Therapist has worked with me at the greenhouse, helps bring real meaning to our students' lives.

I am indebted to the Department of Correction staff, who saw a need for our program and continue to provide support with their greenhouse, land, vehicles and officers. Our program is successful due to the talent and interest of our assigned officers who have fostered an environment conducive to learning. Officers T. Ross and G. Pereira, and later, S. Thomas and E. Guzman went above and beyond duty to "make things happen" on Rikers and create both a family and school atmosphere in the country's largest jail system. Carol Banfield, Farm Director, Leasa Macleash, former Assistant Commissioner of Special Programs, and Harry Bragg, former Farm

Director, have provided a foundation of support for the program. Ms. Mashere Pride-Rawls, former Executive Director of Special Programs, was a constant source of help and always made me laugh.

Katherine Powis, HSNY Librarian has been an invaluable resource for this book, and I am grateful for her work and presence at the Society. A special heartfelt thank you goes to Jennifer Klopp, former Development Director of HSNY, who shaped and edited several versions of this manuscript and raised the funds to move the book forward.

Tom McCarthy, former Department of Correction Assistant Commissioner of Public Affairs who now runs the Department of Correction history website, deserves thanks for all his help securing photographs and sharing his knowledge of Rikers Island. I cannot overstate my gratitude to graphic designer Rita Lascaro for bringing the manuscript to life visually with great skill and patience.

Doing Time in the Garden was initially funded by grants from the William H. Donner Foundation, the Josiah Macy Jr. Foundation and the Robert Wood Johnson Foundation. I am thankful for their support and hope, after five years of work and numerous versions and rewrites, they are pleased with the finished product. I am ultimately grateful to Donna Metty of the Educational Foundation of America for connecting us to our publisher—Lynne Elizabeth and Karen Kearney of New Village Press provided vision, guidance and faith in producing a book that exceeds all expectations.

The GreenHouse and the GreenTeam programs that this book describes would only be concepts were it not for the generous contributions of the foundations and corporations who over the years have enabled us to develop, maintain and expand these programs for men and women both in and out of jail. These supporters are: ADCO Foundation, Altman Foundation, Ambrose Monell Foundation, American Chai Trust, American Express Foundation, Burpee Foundation, City Gardens Club of New York City, Cowles Charitable Trust, Ira W. DeCamp Foundation, William H. Donner Foundation, Dorr Foundation, Jean and Louis Dreyfus Foundation, Educational Foundation of America, Emy & Emil Herzfeld Foundation, Charles A. Frueauff Foundation, Herman Goldman Foundation, Edwin Gould Foundation for Children, William and Mary Greve Foundation,

Hagedorn Fund, Independence Community Foundation, J.M. Kaplan Fund, Louise and Arde Bulova Fund, Richard Lounsbery Foundation, J.P Morgan Chase Foundation, New York Community Trust, Ostgrodd Foundation, Panaphil Foundation, Patrina Foundation, Rhodebeck Charitable Trust, Starr Foundation, United States Environmental Protection Agency and van Ameringen Foundation.

Since GreenHouse began in 1997, many individuals have also contributed either financial or in-kind support to our program. Without them, GreenHouse would not have broken ground as one of the few jail-to-street programs in the vast and tangled U.S. criminal justice system. I thank each one of you who has made this contribution and apologize there is not space to recognize everyone. I would, however, like to mention a few individuals whose help has been invaluable: Ramesh Ahmadi, Sharen and James Benenson, William Burch, the late Jay Chiat, Marcia DaCosta, Rob and Heather Faris, the late George Hecht, Paula Hayes, Dennis Hersch, Julie Houston, Pamela Ito, Michael Jacobson, Millicent Mercer Johnson, Eilleen Jones, Bobbie Margolis, Ross Martin, David Murbach and Erik Pauze of Rockefeller Center, Sharif Nelson, Helen Pratt, Charles M. Royce, Paul and Joyce Scharfer, David Seeler and Bayberry Nursery, Richard Schnall of Rosedale Nursery, Officer Ralph Smith, Catherine Sneed, Jessica Tcherepnine, Francis Torres, Jeremy Travis, Imam Umar of DOC, Edwina Van Gal, Lulu C. Wang, and Kim Wickers.

INDEX

THE HORTICULTURAL SOCIETY OF NEW YORK

FOUNDED IN 1900, the Horticultural Society of New York (HSNY) promotes public understanding of the art and science of horticulture, using the garden as a learning environment. HSNY enhances community life in New York through its library, art gallery, community outreach and educational programs.

HSNY brings horticulture and environmental science to the public schools; offers intensive horticultural education, job training and placement, and aftercare for Rikers Island inmates; and creates high quality gardens in underserved communities around public libraries. HSNY also presents a broad variety of adult education courses, lectures and workshops.

HSNY's circulating and reference library contains nearly 12,000 volumes on horticulture, garden design and garden history, as well as periodicals, nursery, seed and supply catalogs and hundreds of ready-reference resource files, supplemented by access to online information. The library is used by amateur and expert gardeners, professional artists, designers, writers and anyone interested in plants.

The HSNY Art Gallery provides many fine exhibitions, including the annual American Society of Botanical Artists International Juried Exhibition.

For more information:

The Horticultural Society of New York
148 West 37th Street, 13th Floor
New York, NY 10018-6909
phone: 212.757.0915
email: hsny@hsny.org
www.hsny.org

ABOUT THE AUTHOR

JAMES JILER has directed the Horticultural Society of New York's jail-to-street GreenHouse program at Rikers Island since the program's inception in 1997. He previously worked as an urban ecologist in inner-city neighborhoods in Baltimore and New Haven, and with USAID developing an urban green belt in Ahmedabad, India. James also lived in Kathmandu, Nepal, where he served the Ministry of Agriculture establishing ecological farming systems in the Himalayas, lectured about writing at Nepal's graduate university, and guided Himalayan treks for a US based adventure travel company.

James has been a lecturer with the New York Botanical Garden's School of Continuing Education and has appeared on National Public Radio, CBS Sunday Morning Show, Japan TV, WINS Radio, and a recent documentary called the "Healing Gardens" that features his work at Rikers Island. He holds a Masters Degree in Forestry and Social Ecology from Yale University.

James currently resides in New York City. This is his first book.

365.66 J61 INFCW
Jiler, James,
Doing time in the garden :life lessons through prison horticulture /
CENTRAL LIBRARY
04/10
re-conv. 2014